ISSI Scientific Report Series

Volume 18

Series Editor

International Space Science Institute, Bern, Switzerland

The ISSI Scientific Report Series present the results of Working Groups (or Teams) that set out to assemble an expert overview of the latest research methods and observation techniques in a variety of fields in space science and astronomy. The Working Groups are organized by the International Space Science Institute (ISSI) in Bern, Switzerland. ISSI's main task is to contribute to the achievement of a deeper understanding of the results from space-research missions, adding value to those results through multi-disciplinary research in an atmosphere of international cooperation.

Arvind Parmar • Roger-Maurice Bonnet •
Guido De Marchi • Pedro García-Lario •
Erik Kuulkers • Göran Pilbratt •
Celia Sánchez-Fernández • Maria Santos-Lleó •
Norbert Schartel • John Zarnecki

ESA Science Programme Missions

Contributions and Exploitation

Edited by Arvind Parmar

 Springer

Authors

Arvind Parmar
Department of Space and Climate Physics
MSSL/UCL
Dorking, UK

Guido De Marchi
European Space Agency
ESTEC
Noordwijk, The Netherlands

Erik Kuulkers
European Space Agency
ESTEC
Noordwijk, The Netherlands

Celia Sánchez-Fernández
European Space Agency
ESAC
Madrid, Spain

Norbert Schartel
European Space Agency
ESAC
Madrid, Spain

Roger-Maurice Bonnet
Institut d'Astrophysique de Paris
Paris, France

Pedro García-Lario
European Space Agency
ESAC
Madrid, Spain

Göran Pilbratt
Göteborg, Sweden

Maria Santos-Lleó
European Space Agency
ESAC
Madrid, Spain

John Zarnecki
School of Physical Sciences
The Open University
Milton Keynes, UK

Volume Editor

Arvind Parmar
Department of Space and Climate Physics
MSSL/UCL
Dorking, UK

ISSN 2946-1278   ISSN 2946-1286   (electronic)
ISSI Scientific Report Series
ISBN 978-3-031-69003-7   ISBN 978-3-031-69004-4   (eBook)
https://doi.org/10.1007/978-3-031-69004-4

This work was supported by International Space Science Institute.

Cover illustration: M4 mission themes. Exoplanets, plasma physics and the X-ray Universe are the topics chosen by ESA to be considered for the fourth medium-class mission slot, for launch in 2025. Credit: ATG medialab under ESA Standard Licence.

This Springer imprint is published by the registered company Springer Nature Switzerland AG
The registered company address is: Gewerbestrasse 11, 6330 Cham, Switzerland

If disposing of this product, please recycle the paper.

Foreword

The European Space Agency's Science Programme is a major part of ESA's wide-ranging activities that are mandatory for all the 22 Member States. This programme is at the very heart of ESA and is now widely regarded as providing world-leading science. However, that has not always been the case—in the 1970s, the Science programme, although containing some excellent elements, was much more modest in scope and without an overall strategic plan. The situation evolved when it was decided to join the European Launcher Development Organisation (ELDO) and European Space Research Organisation (ESRO) to create ESA. With ESA, Europe demonstrated its ambitions for an independent and strong role in space. Science became a mandatory element for all countries who wanted to be part of ESA. A key industrial return policy and a new strategic approach were implemented within a long-term plan, the Horizon 2000 programme. This was the fruit of a very large consultation process within the scientific community in Europe and beyond. Thus the ideas, and sometimes dreams, of European scientists were identified in all domains of Space Science to fit into a programme with a 20-year horizon, and incorporating the technological needs, thanks to an unprecedented exchange between researchers and industry.

From modest beginnings with COS-B, ESA's first high energy astrophysics observatory launched in 1975 and the GEOS satellite observing the Earth's magnetosphere launched in 1977, ESA continued with pioneering iconic missions such as Giotto with the first close encounter with a comet and Hipparcos providing the first space-based catalogue of the stars. The Scientific Programme then continued at some pace, thanks also to a careful use of the budget allocated and it evolved to encompass most areas of Space Science.

Despite being able to call on only a fraction of the financial resources of the equivalent NASA programme, the ESA Science Programme shows itself to be highly competitive in many areas and in fact to be world-leading in some. From the Sun through many of the planets and primitive bodies of the Solar System to exoplanets, to the energetic and cool universe, there are few significant areas of Space Science where ESA missions have not had a major impact. ESA, supported by a large number of dedicated scientists and engineers, has not been afraid to take the lead with ambitious missions, whether it is Gaia, mapping the positions of stars to unprecedented accuracy or Rosetta orbiting a comet at close quarters for 17 months and even landing a probe on its surface.

These are highly challenging missions which have enabled science to be carried out truly at the cutting edge. ESA has also shown itself to be a willing and reliable partner in projects led by other agencies, for example those from the USA and Japan.

But how can one truly judge the value and effectiveness of such large and complex undertakings? There is probably not a single parameter or criterion which would answer this question. And of course, it depends on what is your definition of value and effectiveness. These would be very different if you were judging from the perspective of technological development or benefit to industry or scientific impact. This study does not aim to address that question. However, it does, perhaps for the first time, attempt to perform various analyses using the large dataset that is now available after more than 40 years of ESA Science Programme activities. It has also attempted to look for variations in performance by Member State, by gender and by age. Although some analyses have previously been performed along some of these lines, these have been on a rather ad-hoc basis, considering specific projects only and over limited time spans. This we believe is the most comprehensive analysis that has been performed to date. While the use of such analyses may be seen as contributing towards a better understanding of effectiveness, they can also contribute towards seeking out in the ESA Science Programme any biases, along the lines of gender and age, should these exist and, if so, whether they be due to conscious or unconscious bias.

Director of the ESA Scientific Programme   Roger-Maurice Bonnet
(1983–2001)

Contents

1 The ESA Science Programme 1
Arvind Parmar, Roger-Maurice Bonnet, and John Zarnecki
References 7

2 ESA Mission Publications and Their Impact 9
Guido De Marchi and Arvind Parmar
    2.1   Introduction 10
    2.2   Publication Libraries 10
    2.3   Overall Publication Numbers and Authors Affiliations 11
    2.4   Publication Metrics 29
    2.5   Cross Mission Publications 45
    2.6   Archival Publications 46
    2.7   Conclusions 66
    References 67

3 Payload Provision to the ESA Science Programme 71
John Zarnecki and Arvind Parmar
    3.1   Introduction 71
    3.2   Methodology 73
    3.3   Payload Contributions 77
    3.4   Payload Complexity and Size 80
            3.4.1   Mission Class 80
            3.4.2   Number of PIs and co-PIs 81
            3.4.3   Co-PI Contributions 81
    3.5   Effect of Weighting Factors 81
    3.6   Payload PI and co-PI Genders 82
    3.7   Conclusions 84
    References 85

4 XMM-Newton Observing Time Proposals 87
Arvind Parmar, Norbert Schartel, and Maria Santos-Lleó
    4.1   Introduction 88

4.2   Science Results   88
4.3   Scientific Productivity   91
4.4   Observing Time   92
4.5   Observation Time Allocation Committee (OTAC)   93
4.6   XMM-Newton User Group   96
4.7   Proposers   97
       4.7.1   Proposers' Institute Countries   97
       4.7.2   Proposers' Gender   100
       4.7.3   Proposers' Academic Age   102
4.8   Proposal Selection   105
       4.8.1   Proposal Selection: Priority A, B or C   105
       4.8.2   Proposal Selection: Priority A or B   112
       4.8.3   Proposal Selection: Outcomes   113
       4.8.4   Proposal Selection: Regional Dependence   115
4.9   Discussion   116
4.10   Further Activities   118
References   119

5 INTEGRAL Observing Time Proposals   123
   Erik Kuulkers, Celia Sánchez-Fernández, and Arvind Parmar
5.1   Introduction   124
5.2   Science Results   125
5.3   Scientific Productivity   127
5.4   Observing Time   128
5.5   Time Allocation Committee (TAC)   131
5.6   INTEGRAL Users Group   134
5.7   Proposers   134
       5.7.1   Proposers' Institute Countries   135
       5.7.2   Proposers' Gender   136
5.8   Proposal Selection   138
       5.8.1   Proposers' Age   140
       5.8.2   Priority A, B and C Proposal Acceptance Rates   140
       5.8.3   Priority A and B Proposal Acceptance Rates   143
       5.8.4   Priority A, B and C Observing Time Acceptance Rates   144
       5.8.5   Priority A and B Observing Time Acceptance Rates   147
5.9   Proposal Selection Summary   148
5.10   Discussion   150
References   151

6 Herschel Observing Time Proposals   155
   Göran Pilbratt, Pedro García-Lario, and Arvind Parmar
6.1   Introduction   156
6.2   Mission Overview   156
6.3   Science Objectives and Results   158
6.4   Science Productivity   163
6.5   Observing Time   165

6.6   Herschel Observing Time Allocation Committee (HOTAC) 166
6.7   Herschel Users' Group (HUG) 167
6.8   Proposers 168
6.9   Proposal Selection 169
6.10   Comparison with the Infrared Space Observatory (ISO) 175
6.11   Discussion 177
References 177

7 Conclusions 183
Arvind Parmar, Roger-Maurice Bonnet, Guido De Marchi,
Erik Kuulkers, Göran Pilbratt, Norbert Schartel, and John Zarnecki
7.1   Introduction 184
7.2   Refereed Publications 184
7.3   Payload Contributions 185
7.4   Publications and Payloads 186
7.5   Observing Time Proposals 187
7.6   Final Considerations 189

Index 191

Introduction

This work started by trying to answer the question "how do you evaluate the scientific performance of the ESA Science Programme's missions?" For many years the decision makers responsible for the content of the ESA Science Programme have been provided with information for each mission including the number of publications, the number of these that are highly cited, the total number of citations, various statistical metrics and the number of unique author names. However, this reporting only provided snapshots of the situation and was not widely distributed. Here we report on a systematic study of these metrics and their evolution with time to provide insights into mission successes and the communities exploiting the data provided by the Science Programme's missions. An obvious highlight is the outstanding success of the Gaia mission with over 1600 refereed publications per year for the last three years—indeed precision astrometry is a field that Europe has made its own. However, it is important to remember that the continuing scientific legacy of "older" missions such as IUE and Hipparcos, both of which have around 5000 annual citations some 25 years after operations ceased.

In addition, it was realized that examining the outcomes of the regular announcements of observing opportunities for ESA's observatory missions, INTEGRAL, Herschel and XMM-Newton could provide further insights. In particular, INTEGRAL and XMM-Newton have been operating for over 20 years allowing the evolution of their user communities to be investigated. The question "can you evaluate how well an ESA Time Allocation Committees have awarded observing time?" was posed. This is a complex issue as what does "well" mean especially as you will never know the outcome if a different observation was to be approved—the missed observation could be the one that leads to a major breakthrough! A way of investigating the performance of a Time Allocation Committee is to search for biases in the outcomes. These could take the form of favouring a particular gender, age group, or countries.

The third pillar of this study is to examine the provision of payload elements for the ESA Science Directorate's missions. Generally, such contributions are directly funded by ESA Members State funding entities and so ESA does not have details of costs; these are anyway difficult to compare between countries. Instead the numbers of principal and co-principal investigators from each of the ESA Member States are used as proxies for the costs in order to investigate the provision of payload elements. A comparison of publications and payload contributions gives insights into how different countries are exploiting the results and opportunities coming from ESA's missions compared to their direct contributions to payloads.

We are not aware of any similar systematic study that covers an interval of 40 years and an evolving community of users and contributors. It is our expectation that the results presented here will be invaluable in helping increase the effectiveness of the Science Programme.

Acronym List

AAS	American Astronomical Society
ACP	Aerosol Collector and Pyrolyser
ADS	Astrophysics Data Service
AGN	Active Galactic Nucleus
AO	Announcement of Opportunity
ALMA	Atacama Large Millimetre/sub-millimetre Array
AWG	Astronomy Working Group
CFHT	Canada-France-Hawaii Telescope
Co-I	Co-Investigator
co-PI	Co-Principal Investigator
CNSA	China National Space Agency
CP	Core Programme
CSG	Centre Spatial Guyanais
DDT	Director's Discretionary Time
D-SCI	ESA Director of Science
ELDO	European Launcher Development Organisation
EPIC	European Photon Imaging Camera
ESA	European Space Agency
ESAC	European Space Astronomy Centre
ESO	European Southern Observatory
ESOC	European Space Operations Centre
ESRO	European Space Research Organisation
ESTEC	European Science and Technology Centre
FIRST	Far Infra-Red and Submillimetre space Telescope
FOV	Field of View
FRBs	Fast Radio Bursts
FWHM	Full-Width at Half Maximum
GBT	Green Bank Telescope
GCMS	Gas Chromatograph Mass Spectrometer
GDP	Gross Domestic Product
GP	General Programme

GRB	Gamma-ray Burst
GSFC	Goddard Space Flight Center
GT	Guaranteed Time
GW	Gravitational Wave
G/AGN	Galaxies and Active Galactic Nuclei
HCSS	Herschel Common Science System
HEASARC	High-Energy Astrophysics Archive Centre
HESS	High Energy Stereoscopic System
HIFI	Heterodyne Instrument for the Far Infrared
HIPE	Herschel Interactive Processing Environment
HOTAC	Herschel Observing Time Allocation Committee
HSA	Herschel Science Archive
HSC	Herschel Science Centre
HST	Hubble Space Telescope
HUG	Herschel Users' Group
IAU	International Astronomical Union
ICC	Instrument Control Centre
IPAC	Infrared Processing and Analysis Center
IREM	INTEGRAL Radiation Environment Monitor
ISDC	INTEGRAL Science Data Centre
ISGRI	INTEGRAL Soft Gamma-Ray Imager
ISOC	INTEGRAL Science Operations Centre
ISSI	International Space Science Institute
ISM	Interstellar Medium
ISM/SF/SS	Interstellar Medium, Star Formation and Solar System Objects
ISO	Infrared Space Observatory
ISOCAM	ISO Infrared Camera
ISOPHOT	ISO Photo-polarimeter
ISWT	INTEGRAL Science Working Team
IBIS	Imager On-Board the INTEGRAL Spacecraft
INTEGRAL	INTErnational Gamma-Ray Astrophysics Laboratory
IUE	International Ultraviolet Explorer
IUG	INTEGRAL Users Group
JAXA	Japanese Aerospace Exploration Agency
JEM-X	Joint European X-ray Monitor
JVLA	Jansky Very Large Array
JWST	James Webb Space Telescope
keV	Kilo Electron Volt
KPI	Key Performance Indicator
kpc	kilo parsec
KP	Key Programme
KPs	Key Programmes
KPGT	Key Programme Guaranteed Time
KPOT	Key Programme Open Time
LWS	Long Wave Spectrometer

MEOR	Mission Extension Operations Review
MeV	Million Electron Volt
MEX	Mars Express Mission
MOC	Mission Operations Centre
MPE	Max Planck Institute for Extraterrestrial Physics
Msec	Millions of Seconds
NASA	National Aeronautics and Space Administration
NHSC	NASA Herschel Science Center
NRAO	National Radio Astronomy Observatory
NUP	NHSC Users' Panel
ObsID	Observation Identifier
OM	Optical Monitor
OT	Open Time
OMC	Optical Monitor Camera
OSA	Off-line Scientific Analysis
OTAC	Observation Time Allocation Committee
PACS	Photodetector Array Camera and Spectrometer
PhD	Doctor of Philosophy Degree
PI	Principal Investigator
PICsIT	Pixellated Imaging Caesium Iodide Telescope
PS	Project Scientist
RAL	Rutherford Appleton Laboratory
RGS	Reflection Grating Spectrometer
SAO	Smithsonian Astrophysical Observatory
SARP	Science Advisory Review Panel
SDP	Science Demonstration Phase
SGS	Science Ground Segment
SMP	Science Management Plan
SOC	Science Operations Centre
SOHO	Solar and Heliospheric Observatory
SPI	Spectrometer on INTEGRAL
SPIRE	Spectral and Photometric Imaging REciever
SRON	Netherlands Institute for Space Research
SSAC	Space Science Advisory Committee
SSC	Senior Science Committee
SSEWG	Solar System and Exploration Working Group
SPC	Science Programme Committee
STScI	Space Telescope Science Institute
S/SE	Stars and Stellar Evolution
SWAS	Submillimeter Wave Astronomy Satellite
SWS	Short Wave Spectrometer
TAC	Time Allocation Committee
TDE	Tidal Disruption Event
TNO	Trans-Neptunian Object
ToO	Target of Opportunity

ULIRGs   Ultra Luminous InfraRed Galaxies
UV   Ultra-Violet
URL   Uniform Resource Locator
VLBA   Very Long Baseline Array
VEX   Venus Express Mission
VLT   Very Large Telescope
WHIM   Warm-Hot Intergalactic Medium
XSA   XMM-Newton Science Archive
XUG   XMM-Newton Users' Group

Chapter 1
The ESA Science Programme

Arvind Parmar, Roger-Maurice Bonnet, and John Zarnecki

Abstract We briefly describe the history of the European Space Agency, the scientific topics covered by the Science Programme and the long-term planning process. The different mission types in the current Cosmic Vision plan are outlined together with the mission selection process. How a mission concept progresses through the various study and project phases, together with typical durations, is also outlined. Finally, the composition of the scientific community who could propose for observing time on ESA's observatories is provided using membership of the International Astronomical Union.

Keywords ESA · European Space Agency · ESA Science Programme · Horizon 2000 · Cosmic Vision · ESA Science Advisory Structure · Mission selection · Mission study phases

The European Space Agency (ESA) is an intergovernmental organisation of 22 Members which was established in 1975 in order to provide for and promote, for exclusively peaceful purposes, cooperation among European States in space research and technology and their space applications, with a view to their being used for scientific purposes and for operational space application systems [1]. ESA had eleven founding Member States: Belgium, Denmark, France, West Germany, Ireland, Italy, the Netherlands, Spain, Sweden, Switzerland, and the United Kingdom. These were joined by Austria in 1984, Norway in 1985, Finland

A. Parmar (✉)
Department of Space and Climate Physics, MSSL/UCL, Dorking, UK
e-mail: arvind.parmar@ucl.ac.uk

R.-M. Bonnet
Institut d'Astrophysique de Paris, Paris, France
e-mail: rmbonnet@issibern.ch

J. Zarnecki
The Open University, Milton Keynes, UK
e-mail: J.C.Zarnecki@open.ac.uk

1

A. Parmar et al., *ESA Science Programme Missions*, ISSI Scientific Report Series 18,
https://doi.org/10.1007/978-3-031-69004-4_1

in 1994, Portugal in 2000, Greece and Luxembourg in 2004, the Czech Republic in 2008, Romania in 2011, Poland in 2012, and Estonia and Hungary in 2015. The history of the predecessor organisations of European Launcher Development Organisation (ELDO) and European Space Research Organisation (ESRO) between 1958 and 1973 is summarised in [2] and the history of ESA between 1973 and 1987 is summarised in [3]. ESA is organised into a number of directorates, with overlapping responsibilities for activities such as human spaceflight, launchers, space science, Earth observation etc. Some activities, such as those conducted by the Science Programme, are mandatory and each Member State contributes pro-rata according to their Gross National Product. Other activities, such as launchers and Earth Observation are optional and Member States can choose whether to participate and their level of contribution.

The topics covered by the Science Programme include astronomy, the Solar System, the Earth's magnetosphere and aspects of fundamental physics. These areas of science overlap with the activities of the ESA Human and Robotic Exploration and Earth Observation Directorates, with which there is close collaboration. Science within ESA is seen as important for inspiring young people towards studying STEM—Science, Technology, Engineering and Mathematical subjects. Initially, missions were chosen on an ad-hoc basis as funding became available and ESA launched Cos-B [4], its first major scientific mission in 1975. In 1978 ESA joined the National Aeronautics and Space Administration (NASA) and the United Kingdom in launching International Ultraviolet Explorer (IUE) [5]. In the 1980s, the Science Programme embraced long-term planning of its programme to help consolidate Europe's position at the forefront of space research and to identify necessary technological developments. The first such plan, Horizon 2000 [6], was prepared in 1984 and was followed by Horizon 2000 Plus [7] in 1994 and Cosmic Vision in 2005 [8]. To put these plans into context, comet-exploring Rosetta [9] and its lander, Philae [10], the Cosmic Microwave Background observer Planck [11], the X-ray observatory XMM-Newton [12] and the Far-Infrared observatory Herschel [13] all began life in Horizon 2000. Gaia [14], Lisa Pathfinder [15] and BepiColombo [16] were all conceived in Horizon 2000 Plus. Cosmic Vision missions are still being realised today: the exoplanet mission CHEOPS [17] launched in 2019, and Solar Orbiter [18] launched in 2020. Jupiter Icy Moons Explorer (JUICE; launched in 2023), Athena and LISA [19] are the Large missions in the Cosmic Vision plan. These latter two missions are scheduled for launch in the 2030s.

The Cosmic Vision programme has four types of mission:

- Large Missions: European led flagships with <20% international collaboration and a cost to ESA of around two annual Science Programme budgets. One such mission every seven to eight years.
- Medium Missions: ESA led, or a contribution to international collaboration. The cost to ESA is around one annual budget. One such mission every three to four years.
- Small, or Fast, Missions: A small mission which shares a launch as a passenger and has a cost to ESA of around 0.3 annual budgets.

• Missions of Opportunity: Missions that are led by other agencies, or ESA Directorates, to which the Science Programme contributes as a junior partner.

A similar process "Voyage 2050" has identified three science themes for the next generation of large class missions—the moons of the giant planets, from temperate exoplanets to the Milky Way and new physical probes of the early Universe. The process also recommended investments in areas of technology so that these themes could become a reality in the second half of this century.

The general model for a Science Directorate mission is that ESA is the mission architect in charge of the space segment development, launch, and in-orbit operations, while the Member States directly fund the payloads and parts of the science operations. International collaboration plays an important role with many of the mission led by the Science Programme having international partners. The Science Programme also contributes to missions led by other agencies such as NASA as a junior partner.

The selection of missions is by competition in a bottom-up process through the Science Advisory Structure (see Fig. 1.1). This allows the European scientific community to influence the direction of the programme through participation in its advisory bodies. These bodies make recommendations to the ESA Director of Science (D-SCI) and their recommendations are independently reported to ESA's Science Programme Committee (SPC)—which has overall authority for the content of the Science Programme. The programme's advisory structure includes the Astronomy Working Group (AWG) and the Solar System and Exploration Working Group (SSEWG) who report to the Space Science Advisory Committee (SSAC) which issues recommendations to the D-SCI and other ESA directors, as appropriate. The membership of the Advisory Structure is public whereas the minutes of their meetings are not. Other working groups in life, Earth and physical sciences may also provide advice to the SSAC. As necessary, independent advisers and ad-hoc advisory groups are often utilised to advise on certain mission proposals or the formulation of planning cycles.

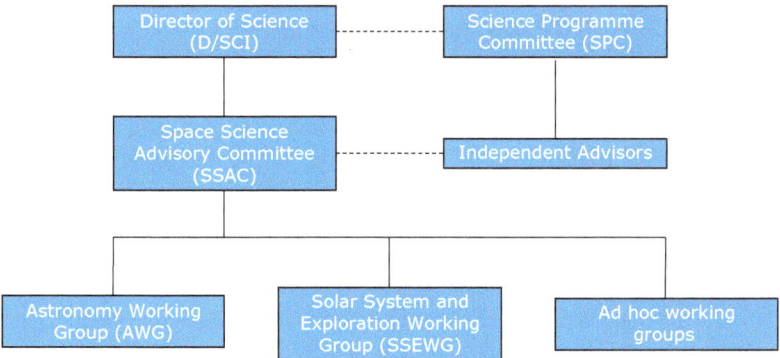

Fig. 1.1 The science programme advisory structure

The process of mission selection starts with the issuing of an Announcement of Opportunity (AO) by D-SCI. This will outline the maximum cost to ESA of the proposed mission, the expected schedule, and the technological readiness levels expected for technologies that the mission may rely on. Information is provided to potential proposers on the capabilities of European launchers, the role of international collaboration, the European tracking ground station network and the support that ESA can provide during any subsequent study activities. The responses to an AO are evaluated by the Advisory Structure and a small number (typically three) mission concepts are chosen for Assessment studies (Phase 0/A). The selection process uses non-conflicted members of the Working Groups who are combined into a single Science Advisory Review Panel (SARP) who evaluate the proposals against a defined set of criteria on science value and feasibility, timeliness, competitiveness and complementarity with other missions. A Senior Science Committee (SSC) consisting primarily of non-conflicted SSAC members then issues reports on each proposal covering the objectives, scientific and programmatic strengths and weaknesses and any final considerations. They also provide a recommendation to D-SCI with the proposals to be selected for study.

The study process starts by using internal ESA facilities to prepare for parallel industrial studies for each concept supported by the proposing scientific teams. Following completion of these feasibility studies, an independent technical and programmatic review is conducted to determine whether the mission concept meets its programmatic boundaries and has a cost to ESA below the maximum allowed. From the concepts that meet these criteria, one is then selected for further study by the Advisory Structure and SPC. The successful concept undergoes a more detailed Definition study. In Phase B1 the mission definition is refined in order to reach readiness for implementation. This includes definition of the spacecraft requirements and instrument interfaces. In addition, any necessary technology development activities are completed. Phase B1 must provide sufficient information to enable Adoption, or final approval, of the mission by the SPC. The mission concept will need to demonstrate that it meets its scientific goals, has a cost to ESA within the given cost cap, has any national funding for the payload and operations in place, together with any international partnership agreements. Following preparatory work in Phase B2, an industrial contract for the flight hardware procurement can then be issued. This normally covers the detailed definition of the mission (Phase C), production and ground testing (Phase D), utilisation for an agreed interval (Phase E) and disposal (Phase F). For a typical Science Directorate medium mission, the interval between initial selection and launch can be 12 or more years (see Table 1.1). The highly competitive nature of the selection process and the rigour imposed by the study process help ensure that only the very best science is enabled by the Science Directorate.

Another important role for the Advisory Structure is to recommend which missions should have their operations extended. Science Missions are often operating beyond their normal lifetimes and are usually extended in 3+3 year cycles, with the second three year interval being subject to a mid-term review. The process starts with an ESA review called the Mission Extension Operations Review (MEOR)

Table 1.1 Typical ESA Science Programme mission activities and durations for each project phase of a Medium Cosmic Vision mission

Project phase	Duration (years)	Activities
0	1	Early concept study: define mission goals, study multiple approaches
A	1	Feasibility study: define functional requirements, choose an approach, analyse alternatives
B1	2	Definition study: define system requirements, complete a preliminary system design allocating functions and defining interfaces
B2/C/D	8	Design, development, test and evaluation: complete detailed system design, fabricate, integrate and test, prepare for launch and operations
E/F		Launch, operations and disposal
Total	12	

which examines the technical outlook, scientific performance, national funding status and takes into count the views of any user group or science team associated with the mission. For missions that successfully pass the review, a mission extension proposal is prepared, an important part of which is the science that is expected to be completed in the mission extension interval. The mission extension proposal is then passed to the Advisory Structure for their recommendation. The way in which this occurs has evolved and now uses an ad-hoc working group consisting mainly of AWG and SSEWG members who are asked to provide a scientific assessment of each of the missions. This is then forwarded to a Senior Committee composed mainly of SSAC members who will be asked to recommend on which missions should be extended, as well as giving a ranking. The Senior Committee's report will be used by D-SCI to produce a proposal to the SPC which includes the affordability to the Science Programme for the missions proposed to be extended for their decision. The process is summarised in Fig. 1.2. As part of their scientific assessment the Advisory Structure may take into account the scientific performance of the missions under review. For this, and other, reasons the Advisory Structure and SPC are regularly informed of the number of publications from each of the missions that the Science Directorate contributes to, either as the mission lead or as a junior partner to a mission led by an external organisation, or other directorates of ESA.

In the chapters on the observatory missions XMM-Newton, INTEGRAL and Herschel, we examine the gender balances, nationalities and ages of the proposers requesting observing time as well as their outcomes. It is useful to compare these with the worldwide population of scientists who could propose for observing time on ESA's Science Programme missions. One way to do this is to use the membership of The IAU. The IAU was founded in 1919 to promote and safeguard the science of astronomy in all its aspects, including research, communication, education and development, through international cooperation. Its members are professional astronomers at the Doctor of Philosophy Degree (PhD) level and above, who are active in professional research, education and outreach in astronomy. The IAU has a total membership of over 12,000 in 93 countries worldwide. It monitors the genders

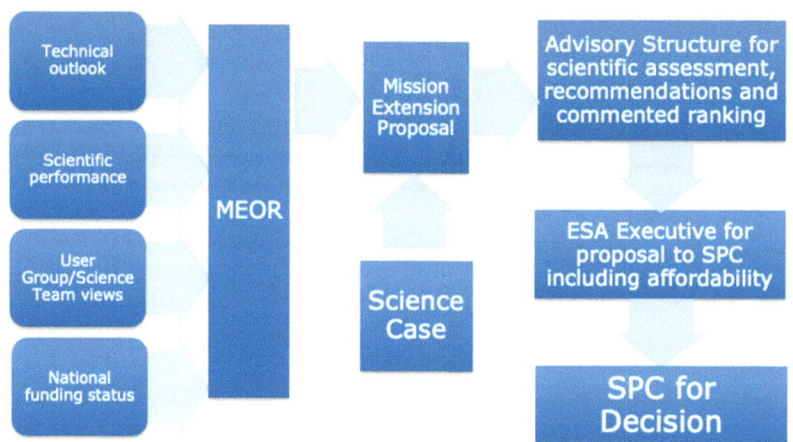

Fig. 1.2 The science programme mission extension process. The MEOR is the ESA mission extension operations review

Table 1.2 The gender distribution with age of the IAU membership of active professional astronomers. Adapted from the IAU website: www.iau.org/administration/membership/individual/distribution/

Age range	Number			Percentage		
	Female	Male	Non-binary or declined to answer	Female	Male	Non-binary or declined to answer
25–30	12	23	3	31.6	60.5	7.9
30–35	189	296	20	37.4	58.6	4.0
35–40	328	563	19	36.0	61.9	2.1
40–45	415	873	23	31.7	66.6	1.8
45–50	371	950	17	27.7	71.0	1.3
50–55	307	1035	13	22.7	76.4	1.0
55–60	264	1149	8	18.6	80.9	0.6
60–65	218	1000	14	17.7	81.2	1.1
65–70	180	906	6	16.5	83.0	0.5
70–75	134	798	6	14.3	85.1	0.6
75–80	115	780	8	12.7	86.4	0.9
80–85	69	610	2	10.1	89.6	0.3
85–90	30	222	1	11.9	87.7	0.4
90–95	5	81	1	5.7	93.1	1.1
95–100	5	24	0	17.2	82.8	0.0

of its members including, recently those who declare non-binary genders. Currently, females comprise 22% of the active membership. However, as Table 1.2 shows, this percentage is strongly age dependent with ~35% of the youngest population being female compared to 10–15% of the population above 70 years old. We note that students are excluded from these statistics, but are active in submitting observing time proposals. We assume that the gender balance of students is similar to that of the 25–35 year old early career scientists.

Acknowledgments We thank Lina Canes of the IAU and Kevin Marvel of the American Astronomical Society for help with finding the past gender balance of the astronomical community. This research was supported by the International Space Science Institute (ISSI) in Bern, through the ISSI Working Group project 'The Scientific Performance of ESA's Science Missions'.

References

1. *Convention for the Establishment of a European Space Agency*. SP–1271(E) (European Space Agency, Paris, 2003)
2. *A History of the European Space Agency*, vol. I. The Story of ESRO and ELDO, 1958 to 1973. SP–1235 (European Space Agency, Paris, 2000)
3. *A History of the European Space Agency*, vol. II. The Story of ESA, 1973 to 1987. SP–1235 (European Space Agency, Paris, 2000)
4. L. Scarsi, K. Bennett, G.F. Bignami, et al., The Cos-B experiment and mission, in *Recent Advances in Gamma-Ray Astronomy*, ed. by R.D. Wills, B. Battrick, vol. 124. ESA Special Publication (European Space Agency, Paris, 1977), p. 3
5. P. Benvenuti, The IUE mission. Memoir. Ital. Astron. Soc. **54**, 359 (1983)
6. *European Space Science – Horizon 2000*. SP–1070 (European Space Agency, Paris, 1984)
7. *Horizon 2000 Plus*. SP–1180 (European Space Agency, Paris, 1994)
8. *Cosmic Vision - Space Science for Europe 2015–2025*. BR–247 (European Space Agency, Paris, 2003)
9. K.H. Glassmeier, H. Boehnhardt, D. Koschny, et al., The Rosetta Mission: flying towards the origin on the solar system. Space Sci. Rev. **128**, 1–21 (Springer 2007)
10. S. McKenna-Lawlor, G. Schwehm, R. Schulz, et al., ESA's comet orbiter rosetta and lander philae, in *Outstanding Problems in Heliophysics: From Coronal Heating to the Edge of the Heliosphere*, ed. by Q. Hu, G.P. Zank. Astronomical Society of the Pacific Conference Series, vol. 484 (Society of the Pacific Astronomical, San Fransisco, 2014). p. 149
11. J.A. Tauber, N. Mandolesi, J.L. Puget, et al., Planck pre-launch status: the planck mission. Astron. Astrophys. **520**, A1 (2010)
12. F. Jansen, D. Lumb, B. Altieri, et al., XMM-Newton observatory. I. The spacecraft and operations. Astron. Astrophys. **365**, L1–L6 (2001)
13. G.L. Pilbratt, J.R. Riedinger, T. Passvogel, et al., Herschel space observatory. An ESA facility for far-infrared and submillimetre astronomy. Astron. Astrophys. **518**, L1 (2010)
14. M.A.C. Perryman, The GAIA mission, in *GAIA Spectroscopy: Science and Technology*, ed. by U. Munari. Astronomical Society of the Pacific Conference Series, vol. 298 (Society of the Pacific Astronomical, San Fransisco, 2003), p. 3
15. P. McNamara, The LISA Pathfinder mission, in *37th COSPAR Scientific Assembly*, vol. 37 (2008), p. 1983
16. J. Benkhoff, G. Murakami, W. Baumjohann, et al., BepiColombo - mission overview and science goals. Space Sci. Rev. **217**(8), 90 (2021)
17. C. Broeg, W. Benz, N. Thomas, et al., The CHEOPS mission. Contrib. Astron. Observator. Skalnate Pleso **43**(3), 498–498 (2014)
18. R. Marsden, B. Fleck, The Solar Orbiter mission, in *34th COSPAR Scientific Assembly*, vol. 34 (2002), p. 222
19. K. Danzmann, LISA mission overview. Adv. Space Res. **25**(6), 1129–1136 (2000)

Chapter 2
ESA Mission Publications and Their Impact

Guido De Marchi and Arvind Parmar

Abstract We examine over 68,000 refereed publications based on data from 25 missions in the ESA Science Programme and 11 additional missions in which ESA is involved as a junior partner. The publications cover the fields of astronomy, planetary science, and heliophysics and are spread over almost 50 years, spanning the period between the year a mission was launched and the end of 2021. We study the number of papers as a function of time and the evolution of several metrics, including citations and other indices. We also investigate the geographical distribution of the authors, and for ESA Member States we correlate the various indices with the level of financial contribution of the individual countries to the ESA Science Programme. We find that in general the involvement of the scientific communities in the various Member States follows the distribution expected from the countries' gross domestic products, with communities in some fields and countries, both large and small, being particularly effective at turning data into scientific discoveries. We also analyse the differences between papers written by investigators directly involved in the provision of the payloads or in the definition of the scientific projects and those written by other scientists not directly involved in the process. We find that the latter, the so-called "archival papers", represent more than 50% of the literature based on data from ESA Space Science missions, and have a similar impact on the literature in the respective fields, as judged by the number of citations. This highlights the importance of sharing and preserving the scientific data produced by the missions.

G. De Marchi (✉)
Directorate of Science, ESA, ESTEC, Noordwijk, The Netherlands
e-mail: gdemarchi@esa.int

A. Parmar
Department of Space and Climate Physics, MSSL/UCL, Dorking, UK
e-mail: arvind.parmar@ucl.ac.uk

Keywords ESA · European Space Agency · ESA Science Programme · Scientific literature · Science productivity · Literature citations · Literature statistics · Literature impact factors · Archival papers

2.1 Introduction

A well-established method of assessing the scientific productivity of an astronomical facility is to perform a quantitative analysis of the publications that have made use of data from that facility. For a reliable comparison across facilities, it is necessary to use publication libraries created with uniform inclusion criteria. In this chapter, we present the analysis of the number and impact of the publications, with the aid of a number of tables and figures. The chapter is structured as follows: Sect. 2.2 addresses the definition and contents of the ESA publication libraries; Sect. 2.3 provides overall publication numbers and discusses the geographical distribution of authors' affiliations; Sect. 2.4 presents standard impact metrics, for the overall ESA Science Programme and for the individual missions, together with an analysis of their temporal evolution; Sect. 2.5 explores papers that use data from more than one mission; Sect. 2.6 discusses the concept of archival publications and provides statistics underlying their contribution to the Space Science literature. Conclusions are presented in Sect. 2.7.

2.2 Publication Libraries

The European Space Agency (ESA) mission publication libraries are found under: https://www.cosmos.esa.int/web/guest/mission-publications. There are 25 libraries covering all ESA-led missions from COS-B [1], launched in 1975, to Solar Orbiter, launched in 2020 (there are also libraries for the JUICE and Euclid missions, which are not covered in this work). The International Ultraviolet Explorer (IUE) was a collaboration between the National Aeronautics and Space Administration (NASA), the Science and Engineering Research Council in the United Kingdom and ESA. It is included in these libraries as an ESA-led mission since ESA took over operational responsibility towards the end of the operational lifetime. In addition, there are 11 libraries for missions where ESA is a junior partner. The libraries are complete beyond the end of 2021, but only publications appearing before the end of 2021 are included in this analysis. With the exception of Hubble Space Telescope (HST), each publication library is maintained by its project scientist, often with the support of Science Operations Centre (SOC) staff. For missions in their archival phase, a "contact scientist" is assigned who is responsible for maintaining the publication library. Space Telescope Science Institute (STScI) staff maintain a list of HST publications which can be accessed via a link from the above Uniform Resource Locator (URL). The libraries make use of the Smithsonian

Astrophysical Observatory (SAO)/NASA Astrophysics Data Service (ADS) which is a digital library for researchers in physics and astronomy operated by the SAO under a NASA grant [2–5]. An important advantage of using ADS is the extensive range of statistical analysis and database selection tools that are provided.

Refereed publications, published after launch, are included in the ESA libraries if they fulfil one or more of the following conditions:

1. Make direct use of data from a mission including from its primary catalogues.
2. Make quantitative predictions of results from a mission.
3. Describe a mission, its instruments, operations, software or calibrations.

Thus, publications that only refer to published results from a mission, and do not fulfil one of the criteria listed above, are not included in the mission publication libraries, but are counted as citations. Although the publications may vary in quality, depth and importance, their number provides an easily quantifiable (although not homogeneous qualitatively) assessment of the productivity of each of the Science Programme missions. Note that a publication will appear in multiple libraries if it makes direct use of data from more that one ESA mission.

Libraries of Doctor of Philosophy Degree (PhD)s from some of the missions are also available under the above web address. The PhD libraries include degrees awarded before launch so that PhDs describing developments prior to launch can be included.

Another very useful resource is the NASA's High-Energy Astrophysics Archive Centre (HEASARC) bibliographical database which includes links to ADS publication libraries for the high-energy astronomy and cosmic microwave background missions supported by the HEASARC (https://heasarc.gsfc.nasa.gov/docs/heasarc/biblio/pubs/).

2.3 Overall Publication Numbers and Authors Affiliations

Table 2.1 and Fig. 2.1 show the number of refereed publications for the ESA-led Science Programme missions for each year in the period considered here, and Table 2.2 and Fig. 2.2 the same for the partner-led missions. For missions launched before 2000, the number of refereed publications before this date are summed. The year of launch and end of operations (if applicable) are given for each mission. These numbers of publications can be obtained from the ESA libraries by selecting publications before the end of 2021 and also selecting on "refereed" as some of the libraries contain a small number of non-refereed publications. It can be seen that:

1. The ESA-led missions with the most publications are XMM-Newton [6], Solar and Heliospheric Observatory (SOHO) [7] and Gaia [8], all with >6000 refereed publications at the end of 2021. Two of these (SOHO and XMM-Newton) are long-lived (>20 years of operations), whereas the Gaia [8] publications in the table span only 8 years.

Table 2.1 The number of refereed publications per year for the ESA-led Science Programme missions using the inclusion criteria given in Sect. 2.2. The missions are arranged in order of launch date. For missions still being operated, only the launch date is given under "Operations". The total number of refereed publications until the end of 2021 is given under "Total Pubs"

Mission	Operations	<2000	2000	2001	2002	2003	2004	2005	2006	2007	2008	2009	2010
COS-B	1975-1982	142	1										
IUE	1978-1996	3713	115	115	98	85	80	95	60	63	60	51	65
Exosat	1983-1986	706	8	3	4			2	1	2	1	2	
Giotto	1985-1992	246	1		3	4			2	2		4	
Hipparcos	1989-1993	714	186	151	144	130	104	96	113	88	87	87	85
Ulysses	1990-2009	898	83	138	39	89	57	56	36	42	52	38	59
ISO	1995-1998	490	180	138	146	129	123	106	67	35	33	76	83
SOHO	1995-	596	286	200	284	299	319	320	273	364	336	322	274
Huygens	1997-2005		7	8	15	7	12	24	21	26	26	16	15
XMM-Newton	1999-		23	90	108	238	347	329	349	361	337	366	365
Cluster	2000-	54	1	38	22	65	126	171	145	123	183	175	193
INTEGRAL	2002-				5	91	42	89	139	112	120	139	129
SMART-1	2003-2006						2	7	10	4	8	10	2
Mars Express	2003-					4	21	49	87	75	83	90	110
Rosetta	2004-2016						23	19	35	59	30	21	41
Venus Express	2005-2014								20	39	72	44	38
Herschel	2009-2013												228
Planck	2009-2013												
PROBA-2	2009-												
Gaia	2013-												
LISA Pathfinder	2015-2017												
ExoMars 2016	2016-												
BepiColombo	2018-												
CHEOPS	2019-												
Solar Orbiter	2020-												

Mission	Operations	2011	2012	2013	2014	2015	2016	2017	2018	2019	2020	2021	Total Pubs
COS-B	1975-1982					2	1						146
IUE	1978-1996	44	49	54	49	55	46	41	39	47	45	36	5105
Exosat	1983-1986		1	2	3	2	1	1	2	1	1		743
Giotto	1985-1992	2		1		1							266
Hipparcos	1989-1993	67	88	71	58	60	54	57	71	62	42	77	2692
Ulysses	1990-2005	34	43	32	34	30	14	36	24	35	33	48	1950
ISO	1995-1998	77	48	41	58	35	26	38	42	28	31	32	2062
SOHO	1995-	298	245	184	213	218	171	210	181	169	230	260	6252
Huygens	1997-2005	8	11	7	5		6	3	1	1	1	2	222
XMM-Newton	1999-	372	326	364	361	337	388	395	397	385	359	366	6963
Cluster	2000-	234	187	174	151	175	159	137	127	108	88	126	2962
INTEGRAL	2002-	93	114	78	101	95	93	105	76	84	84	115	1904
SMART-1	2003-2006	5	6	2	4	2	1	2	4	1			70
Mars Express	2003-	97	82	68	99	101	64	88	98	70	62	116	1464
Rosetta	2004-2016	26	53	32	50	157	188	177	102	131	100	70	1314
Venus Express	2005-2014	53	72	49	40	100	61	46	31	22	25	25	737
Herschel	2009-2013	109	256	323	347	301	305	276	248	234	250	407	3284
Planck	2009-2013	48	68	118	326	327	314	299	297	336	307	302	2742
PROBA-2	2009-	7	6	26	15	11	18	9	12	9	11	13	137
Gaia	2013-			31	37	63	403	881	1633	1692	1663	6403
LISA Pathfinder	2015-2017					3	5	10	8	3	4	33
ExoMars 2016	2016-					9	28	25	20	14	42	138
BepiColombo	2018-							4	11	28	28	71
CHEOPS	2019-								11	13	14	38
Solar Orbiter	2020-									30	84	114

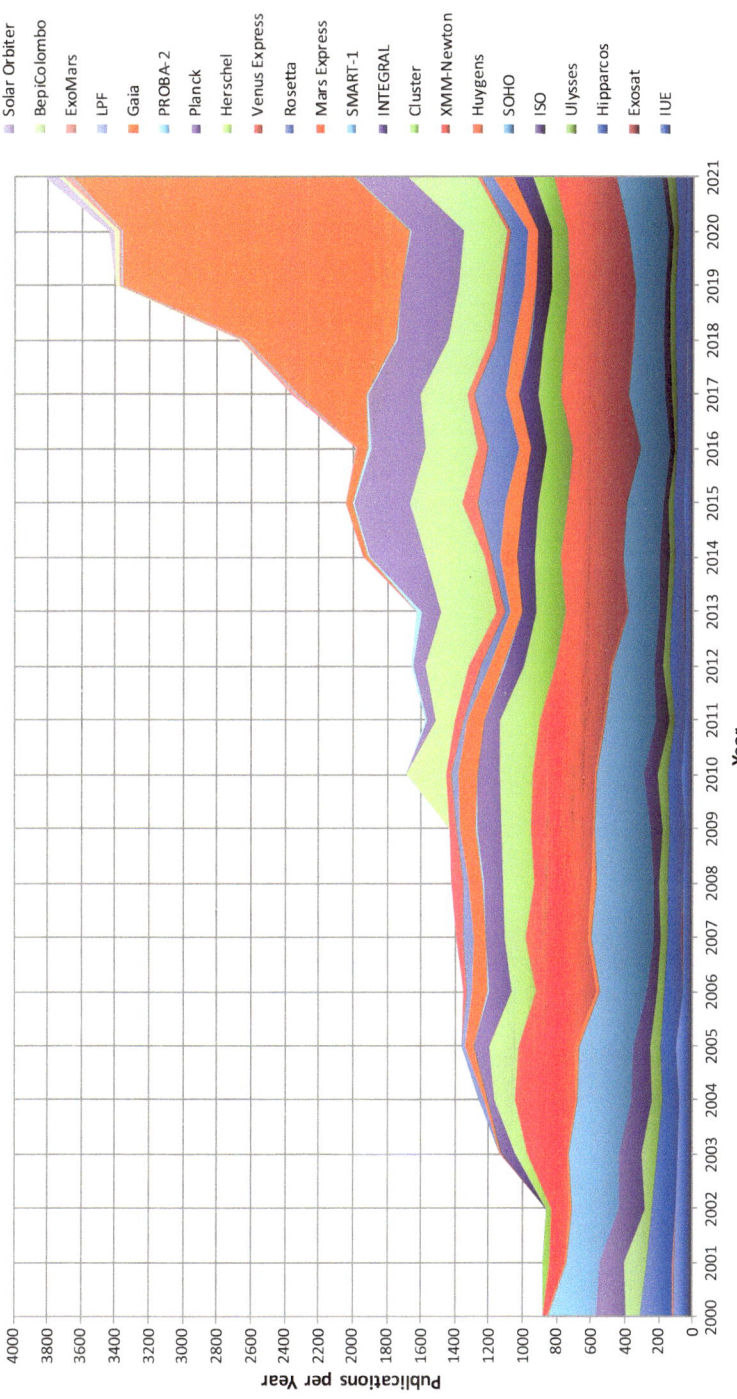

Fig. 2.1 A "waterfall" diagram of the ESA Science Programme Mission refereed publications per year between 2000 and 2021. The missions are arranged in order of launch date with the earliest missions at the bottom. The large contribution from Gaia after 2016 (shown in light brown) is striking

Table 2.2 The number of refereed publications per year for the partner-led Science Programme missions using the inclusion criteria given in Sect. 2.2. The missions are arranged in order of launch date. For missions still being operated, only the launch date is given under "Operations". The total number of refereed publications until the end of 2021 is given under "Total Pubs"

Mission	Operations	Year										
		<2000	2000	2001	2002	2003	2004	2005	2006	2007	2008	2009	2010
HST	1990-	2330	537	560	602	607	611	686	715	733	705	680	734
Cassini	1997-2017	22	1	16	18	11	50	109	149	157	172	201	234
Double Star	2003-2007						1	26	6	12	40	11	14
Suzaku	2005-2015							19	49	78	139	92	
AKARI	2006-2011								1	27	20	19	48
Hinode	2006-									63	104	134	134
CoRoT	2006-2013									5	14	39	44
Chandrayaan-1	2008-2009										3	13	15
IRIS	2013-												
Hitomi	2016-2016												
MICROSCOPE	2016-2018												

Mission	Operations	Year											Total
		2011	2012	2013	2014	2015	2016	2017	2018	2019	2020	2021	Pubs
HST	1990-	789	848	785	820	853	883	914	968	1015	929	1029	19333
Cassini	1997-2017	184	177	134	127	83	97	56	72	71	40	46	2227
Double Star	2003-2007	12	6	4	5	4	4	2	1	3	3	7	161
Suzaku	2005-2015	109	103	95	82	85	97	62	46	38	26	19	1139
AKARI	2006-2011	89	115	94	72	64	69	129	89	99	76	75	1086
Hinode	2006-	133	129	100	96	98	86	82	89	59	63	72	1442
CoRoT	2006-2013	18	21	20	15	15	7	10	12	8	7	3	238
Chandrayaan-1	2008-2009	18	21	20	15	15	7	10	12	8	7	3	164
IRIS	2013-			3	22	63	62	63	68	59	59	50	449
Hitomi	2016-2016						14	5	45	4	4	2	74
MICROSCOPE	2016-2018						2	1	6	5	1	2	17

2. The partner-led mission publications are dominated by HST which shows an increase from ~500 publications per year to ~1000 per year between 2000 and 2021. Cassini [9] also contributes strongly to the partner-led publications.
3. Publications from a mission typically continue for many years after the end of the operational and any (typically 2 to 3 year) post-operational phases.
4. There is a trend of an increase in the overall number of publications from the ESA-led missions with time, particularly between 2016 and 2019 where the large increase in Gaia publications contributed strongly to the total tally.
5. There is a core of long-lived Science Programme missions that have contributed significant numbers of publications for more than 20 years. These include Hipparcos [10], the SOHO [7], Cluster [11] and XMM-Newton [6].

It is important to stress that the missions in the ESA Science Programme cover a wide range of areas, they serve communities with rather different sizes, and have produced data over very different timelines, ranging from the few hours of Huygens, to the more than 30 years of HST. Therefore, it is not possible or meaningful to simply compare the number of scientific papers across missions or try and establish

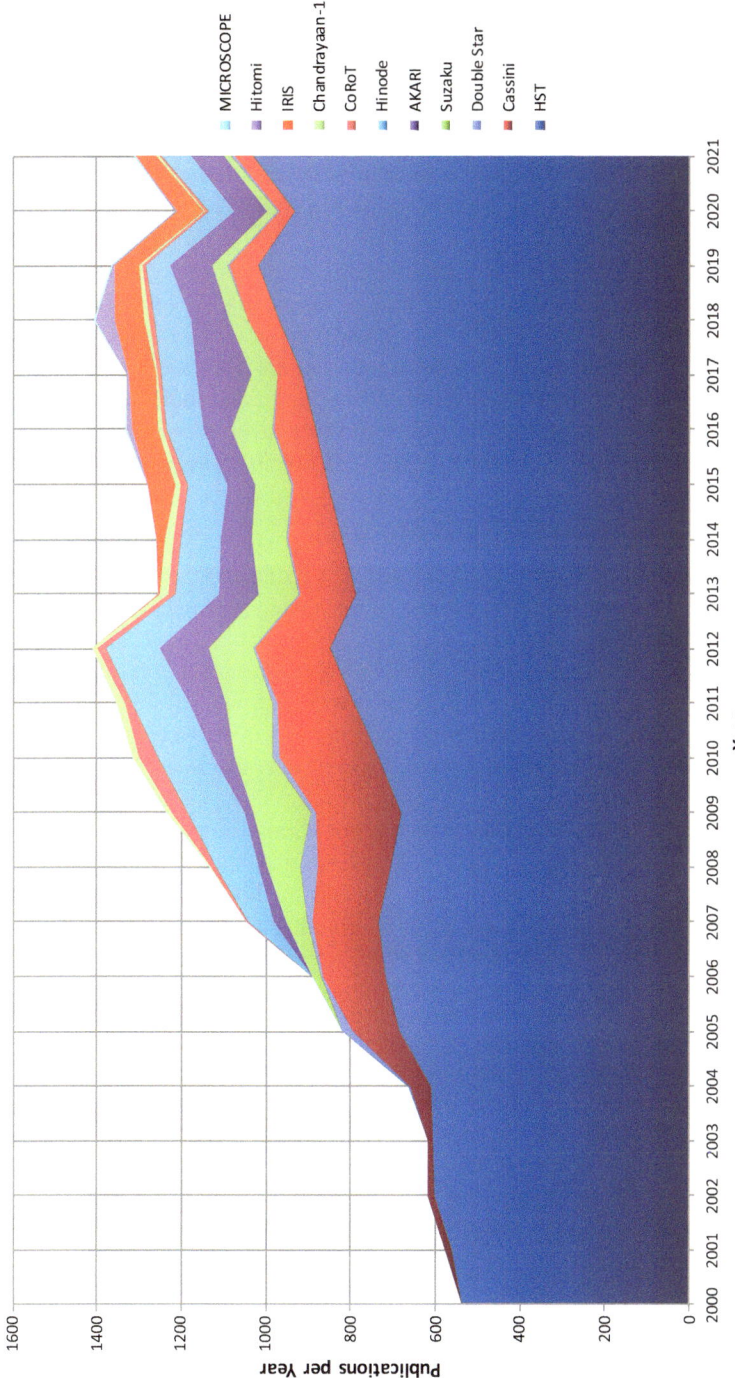

Fig. 2.2 A "waterfall" diagram of the refereed publications per year between 2000 and 2021 for the partner-led missions in the ESA Science Programme. The missions are arranged in order of launch date with the earliest missions at the bottom

whether one is more or less successful than another. Instead, it is useful to explore the evolution of each mission, in terms of number of publications, but also how the scientific communities, in ESA Member States and elsewhere, have taken advantage of the research opportunities offered by the mission data.

We examined the institutes, or universities, in the 22 ESA Member States where the authors of the publications listed in Tables 2.1 and 2.2 are located. If available, these were extracted directly from the publications, or were derived using the names of the author's institute. For publications that were co-authored by a consortium, e.g., the Planck Collaboration, the first authors were assumed to have locations in proportion to those of all the authors listed in the consortium—e.g., a consortium with 50 members with 20 in country A and 30 in country B would have a first author count of 0.4 for country A and 0.6 for country B. Note that the countries used here are the affiliations listed in the publications and are not necessarily the same as the nationalities of the authors. For authors who indicated multiple affiliations, these were distributed pro-rata amongst the countries involved—e.g., if an author indicated affiliations in Greece and Germany, each country was counted as having 0.5 authors.

The authors' affiliations are presented in three ways:

1. Using the countries where the first authors of a publication are affiliated. This can provide an indication of the number of leadership positions for a particular mission in each included country.
2. Using the countries where *all* the authors of a publication are affiliated. This gives an indication of the total contribution of a country to a mission. We caution that this statistic may be affected by a small number of publications with large consortia often with many hundreds of members from other missions or facilities that may not have contributed directly to the specific ESA mission's results.
3. Using the countries of all authors as above, but normalising each publication to have a total of 1.0 authors. So if a publication has ten authors, the countries where each are affiliated would be counted as 0.1. This method prevents the typically few publications from large consortia from dominating the statistics and is our favoured method of investigating the *overall* contributions of different countries to the data exploitation of the missions.

We first present publication statistics in tabular form, using all three methods for ESA-led missions by authors located in the ESA Member States (Tables 2.3, 2.4 and 2.5), for partner-led missions by authors located in the ESA Member States (Tables 2.6, 2.7 and 2.8), and for all missions by authors located in some of the (non-ESA Member State) selected major space powers (Tables 2.9, 2.10 and 2.11). We then examine the implications for the utilisation of the science results from ESA's missions by various countries, which are summarised in Figs. 2.3, 2.4, 2.5, and 2.6.

The author's locations were analysed in the three ways given above. The names of the ESA Member States together with selected major space powers are listed using their ISO 3166 country codes along the top of the tables, while the number of publications are given separately for each mission and country. The results for the

Table 2.3 The number of first authors for the ESA-led mission publications from institutes located in the ESA Member States. Refereed publications that appeared until the end of 2021 and meet the inclusion criteria listed in Sect. 2.2 are included. We indicate with "% Contrib." the percentage average financial contribution of a Member State to the Science Programme evaluated between 2000 and 2021 or between the year that a Member State joined ESA and 2021

Mission	AT	BE	CH	CZ	DE	DK	EE	ES	FI	FR	GB	GR	HU	IE	IT	LU	NL	NO	PL	PT	RO	SE
COS-B	0	0	1	0	23	0	0	0	0	24	13	0	0	0	35	0	42	0	1	0	0	0
IUE	11	51	66	26	12	4	383	169	7	218	542	16	4	6	300	0	110	9	67	7	0	34
Exosat	1	1	2	3	117	8	0	0	4	21	164	3	0	2	94	0	139	0	6	0	0	2
Giotto	0	0	18	0	80	0	0	0	0	28	35	0	2	5	11	0	12	0	0	0	0	1
Hipparcos	37	73	35	37	216	46	1	121	11	262	120	4	10	2	119	0	59	1	39	5	2	43
Ulysses	7	13	41	3	201	2	0	7	9	116	150	31	17	1	63	0	53	4	17	4	5	2
ISO	26	53	8	1	230	10	0	118	24	251	152	3	16	11	118	0	169	0	9	2	1	42
SOHO	51	82	90	43	383	17	1	79	136	339	503	68	16	25	207	0	25	50	77	5	7	13
Huygens	8	1	1	0	18	0	1	12	4	47	10	2	0	0	6	0	9	0	1	1	0	0
XMM-Newton	19	90	103	22	751	14	5	306	30	293	840	70	10	18	936	0	308	0	37	4	0	16
Cluster	94	48	2	54	144	12	1	9	52	178	267	7	10	1	55	0	35	32	12	0	17	175
INTEGRAL	3	12	99	22	186	12	1	84	18	188	93	17	8	14	319	0	60	1	21	2	1	6
SMART-1	0	0	2	0	6	0	0	2	8	2	18	0	0	0	2	0	13	0	0	0	0	2
MEX	5	30	15	12	158	3	0	29	22	213	84	1	7	5	105	0	22	3	5	2	1	72
Rosetta	21	13	71	2	219	1	0	39	9	196	51	2	8	2	128	0	27	18	26	1	2	59
VEX	29	39	7	1	68	0	0	30	14	67	58	0	2	2	28	0	13	0	1	8	2	25
Herschel	21	79	23	2	347	18	1	172	31	318	428	17	31	3	190	0	137	1	14	8	1	58
Planck	6	13	39	3	153	32	3	118	31	213	250	12	8	2	205	0	37	12	10	9	4	13
PROBA-2	4	27	4	0	3	0	0	2	4	6	13	1	0	3	6	0	5	0	0	0	1	1
Gaia	32	69	85	47	475	54	6	307	12	279	535	19	56	15	294	0	150	0	86	21	1	92
LISA PF	0	0	0	0	1	0	0	9	0	0	1	0	0	0	9	0	8	0	0	0	0	0
ExoMars	0	14	4	0	4	0	0	11	0	10	13	0	0	0	10	0	2	0	0	0	0	1
BepiColombo	4	0	3	0	20	0	0	0	1	5	4	0	0	0	16	0	1	1	0	0	0	0
CHEOPS	4	2	7	0	0	0	0	0	0	2	3	0	1	0	1	0	0	0	0	2	0	1
Solar Orbiter	2	4	7	0	9	0	0	9	1	17	15	0	0	0	18	0	3	0	1	0	0	3
Total	385	714	733	278	4195	229	22	1633	428	3293	4362	273	206	117	3275	0	1439	131	430	81	45	661
% MS Pubs	1.68	3.11	3.20	1.21	18.29	1.00	0.10	7.12	1.87	14.36	19.02	1.19	0.90	0.51	14.28	0.00	6.28	0.57	1.88	0.35	0.20	2.88
% Contrib.	2.17	2.72	3.41	0.92	21.16	1.74	0.12	7.22	1.33	14.94	15.41	1.45	0.62	1.03	11.63	0.20	4.42	2.16	2.65	1.17	0.95	2.58
Ratio	0.77	1.14	0.94	1.31	0.86	0.57	0.81	0.99	1.40	0.96	1.23	0.82	1.45	0.50	1.23	0.00	1.42	0.26	0.71	0.30	0.21	1.12

Table 2.4 Total number of authors for the ESA-led mission publications from institutes located in the ESA Member States. Refereed publications that appeared until the end of 2021 and meet the inclusion criteria listed in Sect. 2.2 are included. We indicate with "% Contrib." the percentage average financial contribution of a Member State to the Science Programme evaluated between 2000 and 2021 or between the year that a Member State joined ESA and 2021

Mission	AT	BE	CH	CZ	DE	DK	EE	ES	FI	FR	GB	GR	HU	IE	IT	LU	NL	NO	PL	PT	RO	SE
COS-B	4	1	2	0	154	0	0	6	0	154	45	0	0	0	283	0	211	0	3	0	0	5
IUE	4	1	2	0	154	0	0	6	0	154	45	0	0	0	283	0	211	0	3	0	0	5
Exosat	4	3	7	10	448	26	0	16	12	88	540	4	4	4	414	0	493	2	21	2	3	8
Giotto	0	1	104	0	384	2	0	0	0	123	135	0	4	25	62	0	26	0	0	0	0	7
Hipparcos	140	374	388	170	1210	241	3	587	48	1655	665	20	62	15	759	0	283	8	180	117	9	166
Ulysses	43	72	227	19	1178	24	0	86	37	664	799	133	34	5	483	0	247	14	100	11	20	21
ISO	131	421	56	7	1700	64	1	957	106	2066	1339	45	109	62	844	0	1151	1	62	5	6	277
SOHO	414	537	570	211	2146	118	0	463	678	2095	2243	362	95	146	1506	0	214	222	262	65	61	85
Huygens	65	2	3	0	108	0	0	99	24	292	85	4	1	0	85	0	100	1	2	6	0	1
XMM-Newton	229	433	986	167	7799	299	27	3431	392	4818	6942	423	118	262	9419	0	2360	37	582	61	5	300
Cluster	808	236	35	263	1114	22	0	58	360	1950	2114	32	48	7	301	0	268	287	42	11	88	1121
INTEGRAL	76	133	840	177	3728	247	7	1503	315	2824	2206	79	130	200	5665	1	826	11	394	23	15	198
SMART-1	1	0	24	0	35	0	0	32	54	44	131	0	0	0	16	0	100	0	0	2	0	13
MEX	59	201	138	25	1345	8	0	220	223	1831	630	5	27	40	890	0	192	19	51	15	3	660
Rosetta	163	195	1059	16	3697	13	0	1010	155	2708	530	7	127	13	2664	0	380	66	183	4	4	688
VEX	349	293	44	10	540	3	0	197	102	710	344	6	27	13	405	0	101	0	20	46	8	293
Herschel	256	1322	397	10	5989	310	12	3322	209	7738	8826	281	274	80	4060	0	2216	28	303	127	8	714
Planck	73	133	593	11	2638	762	28	3228	1070	9102	5994	69	74	150	7827	0	710	653	278	90	71	180
PROBA-2	23	241	41	9	75	0	0	14	21	80	82	10	1	26	34	0	72	1	0	0	13	13
Gaia	418	1377	2473	379	7299	1008	43	6716	234	7009	7550	306	666	173	5994	0	1978	30	1192	838	13	994
LISA PF	0	0	105	1	362	0	0	314	0	125	264	0	0	0	409	0	102	0	0	0	0	0
ExoMars	1	242	79	2	25	1	0	155	10	186	125	0	7	0	145	0	32	0	10	0	2	5
BepiColombo	97	11	49	3	250	1	0	29	59	146	96	2	16	4	310	0	32	0	3	3	0	46
CHEOPS	91	48	326	0	89	1	0	75	0	103	118	0	37	1	127	0	51	0	0	64	0	52
Solar Orbiter	76	126	128	127	519	1	0	326	24	708	496	25	3	10	500	0	99	10	36	0	4	118
Total	3604	6679	8924	1728	44497	3213	135	23714	4170	48300	44395	1870	1881	1269	44491	1	12846	1425	3916	1536	333	6134
% MS Pubs	1.36	2.52	3.37	0.65	16.79	1.21	0.05	8.95	1.57	18.22	16.75	0.71	0.71	0.48	16.79	0.00	4.85	0.54	1.48	0.58	0.13	2.31
% Contrib.	2.17	2.72	3.41	0.92	21.16	1.74	0.12	7.22	1.33	14.94	15.41	1.45	0.62	1.03	11.63	0.20	4.42	2.16	2.65	1.17	0.95	2.58
Ratio	0.63	0.93	0.99	0.71	0.79	0.70	0.43	1.24	1.18	1.22	1.09	0.49	1.15	0.47	1.44	0.00	1.10	0.25	0.56	0.50	0.13	0.90

Table 2.5 Number of authors for the ESA-led mission publications from institutes located in the ESA Member States normalised so that each publication contributes 1.0 authors. Refereed publications that appeared until the end of 2021 and meet the inclusion criteria listed in Sect. 2.2 are included. % Contrib. is the percentage average financial contribution of a Member State to the Science Programme evaluated between 2000 and 2021 or between the year that a Member State joined ESA and 2021

Mission	AT	BE	CH	CZ	DE	DK	EE	ES	FI	FR	GB	GR	HU	IE	IT	LU	NL	NO	PL	PT	RO	SE
COS-B	0.0	0.1	0.3	0.0	24.7	0.0	0.0	0.4	0.0	22.3	12.2	0.0	0.0	0.0	37.7	0.0	36.1	0.0	0.9	0.0	0.0	0.0
IUE	14.3	45.8	68.3	19.0	375.4	13.6	4.7	178.2	7.1	199.7	521.9	14.0	3.6	7.5	294.8	0.0	116.5	10.9	58.5	7.4	0.0	35.7
Exosat	1.6	0.4	3.0	0.9	115.8	7.3	0.0	2.6	3.0	22.5	158.3	1.7	0.3	1.6	91.0	0.0	132.2	0.3	6.4	0.4	0.1	2.0
Giotto	0.0	0.2	16.8	0.0	82.3	0.1	0.0	0.0	0.0	25.6	32.5	0.0	1.2	6.9	12.6	0.0	10.6	0.0	0.0	0.0	0.0	1.8
Hipparcos	31.7	70.1	49.1	35.9	225.2	41.8	1.0	119.2	13.2	252.7	122.0	4.9	10.0	2.0	126.2	0.0	54.4	0.8	38.2	6.7	0.9	40.3
Ulysses	6.0	12.1	44.8	2.4	195.6	3.1	0.0	10.5	8.7	109.3	176.9	33.3	11.1	1.5	64.8	0.0	45.8	2.9	17.5	2.6	5.8	2.2
ISO	20.4	46.8	8.4	0.8	238.3	8.9	0.0	122.4	19.6	261.4	141.3	3.4	11.4	7.0	107.1	0.0	174.9	0.1	10.4	0.8	5.9	35.8
SOHO	60.8	85.6	79.8	41.9	416.8	19.6	0.3	70.4	123.0	363.4	490.0	61.1	19.4	24.4	199.5	0.0	21.7	47.6	65.9	5.1	0.0	13.0
Huygens	8.1	0.5	1.0	0.0	17.4	0.0	0.0	13.7	3.4	43.1	11.2	1.3	0.2	0.0	8.9	0.0	9.9	0.0	0.3	1.3	0.0	0.3
XMM-Newton	21.3	73.7	96.0	18.6	701.0	16.9	2.4	312.8	26.6	317.7	833.1	62.2	7.3	15.4	908.6	0.0	299.3	0.9	41.3	4.2	0.5	18.3
Cluster	100.1	44.3	3.9	46.0	158.3	2.2	0.0	7.0	46.8	253.6	305.3	5.4	7.0	1.5	50.5	0.0	34.9	33.0	8.2	1.3	13.4	166.0
INTEGRAL	3.4	9.8	88.4	23.6	179.5	15.1	1.1	85.9	19.2	185.9	107.4	14.3	5.4	11.8	298.1	0.1	57.1	0.4	22.7	1.4	0.5	6.9
SMART-1	0.1	0.0	2.3	0.0	4.3	0.0	0.0	2.5	6.6	4.3	17.2	0.0	0.0	0.0	2.5	0.0	13.1	0.0	0.0	0.1	0.0	2.1
MEX	4.8	25.7	13.6	4.9	167.8	1.1	0.0	24.8	19.9	238.2	80.4	0.8	7.1	3.8	105.5	0.0	23.2	3.2	5.0	2.3	1.1	68.6
Rosetta	15.0	13.4	64.4	2.4	233.4	1.0	0.0	48.9	12.8	192.1	52.3	0.7	9.1	1.1	120.4	0.0	28.8	13.2	26.4	0.8	0.2	54.8
VEX	38.7	33.2	4.4	1.1	65.2	0.5	0.0	23.9	11.3	85.4	48.5	0.3	2.3	0.7	35.4	0.0	12.9	0.4	1.5	5.2	0.9	29.7
Herschel	20.5	71.4	16.8	1.7	361.5	15.4	0.6	174.7	20.9	356.4	396.8	18.8	21.3	3.6	197.8	0.0	125.0	0.4	17.0	6.9	1.2	49.5
Planck	8.0	13.5	39.2	2.5	145.2	31.6	4.9	124.1	25.5	245.3	257.7	13.3	7.0	2.1	222.2	0.0	37.7	15.4	14.5	11.0	3.4	12.6
PROBA-2	3.6	31.1	3.7	1.4	3.9	0.0	0.0	1.9	2.6	9.7	10.2	0.7	0.1	3.1	5.7	0.0	2.9	0.0	0.0	0.0	0.9	0.9
Gaia	29.8	73.2	85.1	43.8	485.6	52.7	5.5	317.2	10.9	315.5	514.2	15.3	43.5	11.9	289.9	0.0	144.3	1.0	86.6	27.0	0.5	76.6
LISA PF	0.5	0.5	1.2	0.0	5.9	0.1	0.0	3.6	0.0	1.4	3.2	0.0	0.0	0.0	11.1	0.0	1.2	0.0	0.2	0.0	0.0	0.0
ExoMars	0.5	13.2	4.3	0.1	3.9	0.1	0.0	10.3	0.2	13.1	11.6	0.0	0.2	0.0	8.7	0.0	2.5	0.0	0.2	0.0	0.0	0.6
BepiColombo	3.4	0.5	1.8	0.1	20.1	0.0	0.0	1.1	1.4	5.7	3.2	0.0	0.3	0.0	14.1	0.0	1.1	0.1	0.3	0.0	0.0	1.3
CHEOPS	3.3	0.6	6.1	0.0	1.0	0.0	0.0	0.8	0.0	2.0	1.8	0.0	0.4	0.1	1.5	0.0	0.5	0.0	0.0	1.9	0.0	0.7
Solar Orbiter	2.7	3.8	5.1	2.6	12.7	0.0	0.0	8.5	0.8	16.1	15.9	0.3	0.2	0.1	13.7	0.0	2.1	0.7	1.2	0.0	0.1	2.9
Total	398.2	668.9	707.7	249.6	4240.5	231.0	20.5	1665.5	383.4	3542.6	4325.0	251.9	168.1	106.3	3228.3	0.1	1389.0	131.0	423.0	86.4	36.3	622.5
% MS Pubs	1.74	2.92	3.09	1.09	18.54	1.01	0.09	7.28	1.68	15.49	18.91	1.10	0.73	0.46	14.11	0.00	6.07	0.57	1.85	0.38	0.16	2.72
% Contrib.	2.17	2.72	3.41	0.92	21.16	1.74	0.12	7.22	1.33	14.94	15.41	1.45	0.62	1.03	11.63	0.20	4.42	2.16	2.65	1.17	0.95	2.58
Ratio	0.80	1.07	0.91	1.18	0.88	0.58	0.75	1.01	1.26	1.04	1.23	0.76	1.19	0.45	1.21	0.00	1.37	0.27	0.70	0.32	0.17	1.05

Table 2.6 Number of first authors for the partner-led mission publications from institutes located in the ESA Member States. Refereed publications that appeared until the end of 2021 and meet the inclusion criteria listed in Sect. 2.2 are included

Mission	AT	BE	CH	CZ	DE	DK	EE	ES	FI	FR	GB	GR	HU	IE	IT	LU	NL	NO	PL	PT	RO	SE
HST	53	108	196	13	1314	109	8	653	32	673	1567	35	15	39	1042	0	481	5	75	21	3	200
Cassini	12	29	3	10	142	0	0	29	2	174	254	14	8	0	58	0	6	3	2	1	0	36
Double Star	8	0	1	3	5	0	0	0	2	10	26	0	0	4	3	0	3	0	0	0	0	0
Suzaku	0	3	6	2	34	0	1	16	4	13	57	3	2	3	54	0	19	0	7	0	0	2
AKARI	6	26	1	6	54	3	0	40	6	38	69	6	15	1	21	0	12	0	28	0	0	4
Hinode	16	10	12	28	98	2	0	45	1	50	167	17	0	3	45	0	4	15	4	1	1	9
CoRoT	9	23	4	2	22	1	0	16	0	57	10	0	6	0	22	0	9	0	1	2	0	0
Chandrayaan-1	0	0	6	0	4	0	0	0	1	2	15	0	0	0	1	0	4	0	0	0	0	10
IRIS	0	1	12	15	14	0	0	4	0	11	41	4	0	0	11	0	1	38	5	0	0	9
Hitomi	0	0	0	0	1	0	0	0	0	0	2	0	0	4	1	0	2	0	0	0	0	0
MICROSCOPE	0	0	0	0	0	0	0	0	0	15	0	0	0	0	1	0	0	0	0	1	0	0
Total	104	200	241	79	1689	115	9	803	48	1043	2208	79	46	54	1259	0	541	61	122	26	4	270

Table 2.7 Total number of authors for the partner-led mission publications from institutes located in the ESA Member States. Refereed publications that appeared until the end of 2021 and meet the inclusion criteria listed in Sect. 2.2 are included

Mission	AT	BE	CH	CZ	DE	DK	EE	ES	FI	FR	GB	GR	HU	IE	IT	LU	NL	NO	PL	PT	RO	SE
HST	467	991	2452	167	13464	1500	28	6179	383	8904	13991	344	171	260	12223	0	4276	97	708	323	22	1643
Cassini	76	188	10	49	1018	0	0	166	48	1481	1959	111	50	4	589	1	72	25	10	18	1	319
Double Star	134	1	1	12	73	1	0	1	9	166	313	0	0	26	17	0	48	4	1	0	2	19
Suzaku	18	8	114	10	727	33	2	373	44	301	518	10	7	41	799	0	164	1	97	5	1	41
AKARI	46	217	70	27	569	16	0	437	33	642	807	37	116	9	304	0	180	3	222	13	1	59
Hinode	92	69	56	115	468	4	0	200	11	297	767	63	24	13	269	0	16	119	40	1	4	40
CoRoT	181	285	169	5	550	21	0	312	0	1734	175	0	43	0	192	0	155	1	12	30	0	13
Chandrayaan	0	1	45	0	33	0	0	3	23	30	131	0	0	1	5	0	16	1	1	0	0	90
IRIS	1	21	55	70	119	1	0	37	0	84	219	11	10	3	73	0	4	276	17	0	3	47
Hitomi	0	0	49	13	21	0	0	30	0	78	73	0	16	28	15	0	186	0	14	0	0	13
MICROSCOPE	0	0	0	0	18	0	0	1	0	120	3	0	0	0	2	0	2	0	0	1	0	0
Total	1015	1781	3021	468	17060	1576	30	7739	551	13837	18956	576	437	385	14488	1	5119	527	1122	391	33	2284

Table 2.8 Number of authors for the partner-led mission publications from institutes located in the ESA Member States normalised so that each publication contributes 1.0 authors. Refereed publications that appeared until the end of 2021 and meet the inclusion criteria listed in Sect. 2.2 are included

Mission	AT	BE	CH	CZ	DE	DK	EE	ES	FI	FR	GB
HST	46.7	111.6	186.7	14.5	1276.1	100.5	6.4	620.3	39.1	747.7	1523.4
Cassini	9.9	25.3	2.0	6.6	128.5	0.0	0.0	25.0	5.4	177.2	257.7
Double Star	10.2	0.1	0.5	1.4	6.4	0.1	0.0	0.3	1.0	16.9	29.7
Suzaku	0.2	2.2	7.8	1.0	47.2	1.8	0.7	16.2	3.4	17.9	63.3
AKARI	3.5	20.0	3.6	3.8	52.9	1.6	0.0	35.0	2.7	44.0	60.2
Hinode	15.9	10.1	10.7	22.2	97.6	0.8	0.0	48.4	1.2	50.3	158.3
CoRoT	10.5	15.9	4.7	1.0	24.2	1.3	0.0	19.0	0.0	67.3	9.1
Chandrayaan-1	0.0	0.0	4.5	0.0	4.8	0.0	0.0	0.3	1.7	3.4	13.2
IRIS	0.1	2.8	9.9	13.5	17.4	0.2	0.0	6.7	0.0	14.0	43.3
Hitomi	0.0	0.0	0.3	0.0	0.1	0.0	0.0	0.1	0.0	0.8	1.8
MICROSCOPE	0.0	0.0	0.0	0.0	1.4	0.0	0.0	0.5	0.0	14.0	0.2
Total	96.9	188.1	230.8	64.0	1656.7	106.2	7.1	771.9	54.6	1153.5	2160.2

Mission	GR	HU	IE	IT	LU	NL	NO	PL	PT	RO	SE
HST	33.7	13.8	31.7	1033.6	0.0	435.4	7.7	73.0	21.5	1.6	180.3
Cassini	11.2	6.7	0.6	56.5	0.1	7.9	3.0	2.8	1.4	0.1	35.2
Double Star	0.0	0.0	2.1	2.2	0.0	3.7	0.3	0.1	0.0	0.1	1.5
Suzaku	1.9	0.4	2.9	56.1	0.0	18.1	0.0	5.8	0.7	0.0	1.9
AKARI	4.1	10.5	0.8	23.5	0.0	15.1	0.2	21.4	0.9	0.1	4.3
Hinode	15.3	1.7	3.1	40.2	0.0	2.8	19.2	5.3	0.1	0.4	8.6
CoRoT	0.0	4.3	0.0	15.4	0.0	7.0	0.0	0.6	1.9	0.0	0.5
Chandrayaan-1	0.0	0.0	0.1	1.2	0.0	2.6	0.1	0.1	0.0	0.0	11.3
IRIS	2.6	0.8	0.5	10.3	0.0	1.0	40.0	2.8	0.0	0.2	9.7
Hitomi	0.0	0.1	0.1	0.1	0.0	1.8	0.0	0.1	0.0	0.0	0.1
MICROSCOPE	0.0	0.0	0.0	1.0	0.0	0.1	0.0	0.0	0.5	0.0	0.0
Total	68.7	38.3	41.9	1239.9	0.1	495.3	70.7	112.0	27.0	2.5	253.3

first authors of the publications from the ESA-led missions (Table 2.3) are shown graphically in Figs. 2.3, 2.4, and 2.5.

Figure 2.3 displays the percentage of publications from each ESA Member State with respect to the total from all Member States and it reveals that the bulk of the publications are associated with first authors located in Germany, France, United Kingdom, and Italy. To better understand any trends or correlations, we also calculated the average financial contributions to the ESA Science Programme by each Member State between 2000 and 2021. For Member States that joined ESA after 2000 (see Chap. 1), the average values from the year of joining to 2021 were

Table 2.9 Number of first authors for publications using the inclusion criteria given in Sect. 2.2 from scientists located in (non-ESA Member State) major space powers. The "Other" column is the number of authors who are not located in an ESA Member State nor one of the countries listed in this table

Mission	AR	AU	BR	CA	CH	CL	IL	IN	JA	KR	MX	RU	US	Other
COS-B	0	0	0	0	0	13	0	0	0	0	0	3	14	5
IUE	47	34	39	98	24	32	19	51	40	37	59	31	1953	139
Exosat	1	3	0	4	1	2	0	36	6	1	0	3	101	7
Giotto	0	0	0	0	0	0	9	0	0	0	0	4	58	1
Hipparcos	31	48	41	59	51	71	7	18	28	16	19	162	530	250
Ulysses	2	9	6	8	0	43	0	7	28	3	6	55	876	61
ISO	4	11	13	35	15	39	7	19	67	18	16	4	411	43
SOHO	45	35	44	13	2	498	2	275	127	101	26	235	2175	194
Huygens	0	1	1	1	2	0	3	0	2	0	2	2	79	1
XMM-Newton	38	61	31	121	48	229	30	145	280	28	36	83	1621	226
Cluster	1	12	10	27	1	377	3	43	79	14	11	174	663	54
INTEGRAL	6	15	8	17	6	85	3	33	54	4	6	161	260	55
SMART-1	0	0	0	1	0	6	0	6	2	0	0	2	1	2
MEX	2	6	2	21	2	37	0	29	13	1	5	64	410	17
Rosetta	0	0	3	2	2	10	7	5	13	3	3	32	205	34
VEX	2	7	1	4	0	48	3	4	44	0	4	41	145	6
Herschel	22	45	13	96	54	114	7	60	74	23	25	21	729	65
Planck	4	32	49	77	30	215	13	114	83	43	10	34	532	139
PROBA-2	2	0	1	0	0	4	0	5	4	1	0	12	20	4
Gaia	38	168	125	163	192	413	30	112	116	66	49	176	1683	322
LISA PF	0	0	0	0	0	0	0	0	0	0	0	0	5	0
ExoMars 2016	0	1	0	5	0	0	0	1	1	0	0	19	12	4
BepiColombo	0	0	0	0	0	0	0	0	8	0	0	2	1	2
CHEOPS	0	0	0	0	0	0	0	0	0	0	0	0	1	1
Solar Orbiter	0	0	0	0	0	5	0	0	0	1	0	1	18	2
HST	78	406	191	418	264	363	93	146	351	138	162	212	8554	386
Cassini	7	7	6	9	0	19	4	9	32	9	6	11	1148	17
Double Star	0	1	0	0	0	73	0	0	4	0	0	3	16	0
Suzaku	2	2	1	13	0	31	0	39	491	2	2	9	271	33
AKARI	5	18	13	12	16	36	0	20	228	67	7	14	177	64
Hinode	6	6	3	0	0	119	1	67	217	32	0	13	402	27
CoRoT	0	0	18	2	3	1	6	1	0	0	0	3	8	8
Chandrayaan-1	0	0	0	2	0	10	0	56	3	0	0	1	39	8
IRIS	0	1	1	0	0	100	0	24	23	6	0	2	117	4
Hitomi	0	1	0	1	0	0	0	0	34	0	0	1	19	1
MICROSCOPE	0	0	0	0	0	0	0	0	0	0	0	0	0	0

Table 2.10 Total number of authors for publications using the inclusion criteria given in Sect. 2.2 from scientists located in (non-ESA Member State) major space powers. The "Other" column is the number of authors who are not located in an ESA Member State nor one of the countries listed in this table

Mission	AR	AU	BR	CA	CL	CN	IL	IN	JA	KR	MX	RU	US	Other
COS-B	0	7	0	3	0	41	0	0	6	0	0	20	118	21
IUE	164	208	169	489	257	169	81	153	222	118	212	156	8529	602
Exosat	1	24	5	10	19	7	0	97	36	2	0	9	461	50
Giotto	0	0	0	2	0	0	21	2	3	0	0	25	333	4
Hipparcos	92	358	221	305	310	317	25	89	277	105	86	477	3211	1061
Ulysses	13	61	23	50	6	167	6	40	183	17	18	279	4555	248
ISO	34	99	65	256	128	136	58	83	460	79	59	29	3277	235
SOHO	233	140	193	61	7	2518	9	890	742	561	131	1071	10460	904
Huygens	2	7	3	6	8	0	13	1	15	0	6	7	404	19
XMM-Newton	225	1242	377	920	869	1415	215	724	2877	195	448	768	17932	2177
Cluster	23	53	47	134	7	2283	15	140	501	72	36	827	4596	244
INTEGRAL	156	730	156	173	184	858	60	484	1143	175	178	1443	8067	991
SMART-1	0	1	0	1	0	15	0	40	20	0	0	6	26	9
MEX	2	30	8	101	7	184	0	95	105	8	14	378	3098	109
Rosetta	3	6	16	10	21	66	23	38	111	11	10	109	2270	420
VEX	11	31	6	22	1	276	8	20	335	0	15	286	975	48
Herschel	92	882	120	2043	1142	1145	78	360	1435	420	356	170	12764	1195
Planck	14	598	425	1762	487	1071	49	566	903	424	109	520	13665	1409
PROBA-2	11	6	3	2	0	21	0	22	28	7	1	49	163	22
Gaia	283	2936	1404	1571	3094	3244	333	711	2108	710	480	920	25023	3050
LISA PF	0	2	0	0	0	0	0	1	0	0	0	0	152	0
ExoMars	0	5	0	31	0	0	0	8	5	0	0	320	170	46
BepiColombo	2	1	0	0	0	3	0	2	164	0	0	33	108	10
CHEOPS	0	6	0	2	9	0	1	0	0	0	0	1	81	4
Solar Orbiter	2	1	0	3	0	29	0	0	5	12	0	12	606	12
HST	488	3849	1243	3485	3893	2693	998	883	4659	998	1149	1113	76189	3679
Cassini	28	40	30	44	3	84	20	26	157	32	19	71	8746	99
Double Star	0	1	1	8	0	482	0	0	27	0	0	20	182	22
Suzaku	9	36	3	75	40	169	6	151	4343	17	28	66	2558	269
AKARI	19	169	62	187	255	196	3	90	2384	518	74	80	2112	536
Hinode	46	24	20	0	1	568	5	222	1299	199	1	70	2339	118
CoRoT	4	15	187	23	26	5	98	6	4	0	3	9	180	43
Chandrayaan-1	0	0	0	14	0	63	0	316	50	1	0	12	390	39
IRIS	2	6	4	0	7	453	1	64	112	53	0	14	910	24
Hitomi	0	5	0	56	0	0	0	0	2433	1	0	2	1097	5
MICROSCOPE	0	0	0	0	0	0	0	0	0	0	0	0	3	1

Table 2.11 Number of authors for publications using the inclusion criteria given in Sect. 2.2 from scientists located in (non-ESA Member State) major space powers, normalised so that each publication contributes 1.0 authors. The "Other" column is the number of authors who are not located in an ESA Member State nor one of the countries listed in this table

Mission	AR	AU	BR	CA	CL	CN	IL	IN	JA	KR	MX	RU	US	Other
COS-B	0.0	0.8	0.0	0.6	0.0	11.6	0.0	0.0	0.0	0.0	0.0	3.3	15.6	5.2
IUE	43.9	32.3	37.3	104.5	31.6	30.3	16.5	41.6	43.2	32.0	50.0	32.0	2013	139
Exosat	0.3	4.8	1.0	2.9	3.3	1.6	0.0	35.4	7.3	1.0	0.0	3.8	109	8.8
Giotto	0.0	0.0	0.0	0.5	0.0	0.0	6.4	0.2	0.7	0.0	0.0	3.8	59.0	0.9
Hipparcos	33.7	46.0	36.0	60.5	54.5	66.8	6.4	19.3	31.8	15.8	19.0	157	545	240
Ulysses	1.9	7.7	4.9	10.7	0.2	39.1	0.2	9.4	29.4	3.0	4.9	52.3	868	58.6
ISO	4.8	12.1	11.4	31.9	13.9	33.4	9.2	15.4	68.0	16.7	13.9	5.2	450	36.2
SOHO	39.9	32.4	42.4	15.1	2.1	460	2.0	240	143	99.0	23.2	231	2270	176
Huygens	0.1	0.8	1.1	0.9	1.3	0.0	3.2	0.3	1.8	0.0	0.9	2.7	76.9	0.6
XMM-Newton	33.6	82.2	33.2	107	52.7	208	26.7	136	282	21.3	44.5	94.0	1759	211
Cluster	3.8	10.5	8.7	25.1	1.4	298	3.7	38.2	82.1	11.7	6.1	147	696	39.8
INTEGRAL	7.5	16.7	8.0	16.3	9.3	80.6	5.6	30.3	52.6	3.2	6.2	157.4	289	59.1
SMART-1	0.0	0.1	0.0	1.0	0.0	6.2	0.0	5.3	1.9	0.0	0.0	0.9	2.7	1.1
MEX	0.4	5.4	2.0	17.6	1.0	32.7	0.0	26.0	12.7	1.3	3.4	54.7	425	18.0
Rosetta	0.5	0.8	2.6	1.2	1.7	10.0	6.6	5.6	15.1	2.7	1.4	27.7	210	32.3
VEX	1.9	5.0	1.0	3.3	0.2	35.7	2.3	4.2	39.1	0.0	2.8	40.1	155	5.2
Herschel	17.1	53.5	9.4	94.0	68.1	105.2	5.9	47.5	83.8	24.5	27.5	22.2	761	61.3
Planck	3.7	30.9	52.6	79.6	29.2	200.2	7.8	114	85.8	43.3	14.5	42.9	599	137
PROBA-2	1.2	0.2	0.4	0.6	0.0	2.5	0.0	3.6	3.8	1.3	0.0	8.5	22.6	3.9
Gaia	40.0	185	111	161	213	379	28.7	102	129	58.7	53.7	177	1747	305
LPF	0.0	0.5	0.0	0.0	0.0	0.0	0.0	0.1	0.0	0.0	0.0	0.0	3.7	0.0
ExoMars	0.0	0.3	0.0	3.0	0.0	0.0	0.0	1.5	0.2	0.0	0.0	19.3	12.2	4.0
BepiColombo	0.0	0.0	0.0	0.0	0.0	0.0	0.0	0.1	6.6	0.0	0.0	2.1	2.6	1.0
CHEOPS	0.0	0.0	0.0	0.0	0.1	0.0	0.0	0.0	0.0	0.0	0.0	0.2	0.9	0.1
Solar Orbiter	0.2	0.1	0.0	0.0	0.0	3.2	0.0	0.0	0.4	0.2	0.0	0.4	18.8	0.2
HST	64.5	414	168	423	335	317	94.2	129	361	123	176	202	8798	338
Cassini	5.4	6.3	6.2	6.0	0.6	12.5	4.2	7.9	24.6	6.4	6.4	13.9	1182	16.4
DoubleStar	0.0	0.0	0.1	0.6	0.0	55.5	0.0	0.0	2.5	0.0	0.0	1.4	19.4	1.9
Suzaku	0.7	1.9	0.3	11.6	2.3	27.3	0.5	39.5	470	2.6	2.8	9.4	271	30.5
AKARI	3.3	18.3	11.6	11.9	20.4	35.8	0.2	18.5	245	62.9	7.0	15.2	191	60.9
Hinode	7.0	4.1	2.4	0.0	0.3	110.8	0.7	58.0	218	30.3	0.3	14.6	438	23.1
CoRoT	0.2	0.7	14.3	2.5	2.4	1.0	5.0	0.3	0.5	0.0	0.1	2.5	9.0	7.8
Chandrayaan-1	0.0	0.0	0.0	2.4	0.0	7.8	0.0	52.6	5.1	0.1	0.0	1.8	43.8	5.8
IRIS	0.2	1.3	0.7	0.0	0.2	87.9	0.2	17.3	18.0	7.5	0.0	2.2	125	5.4
Hitomi	0.0	1.0	0.0	0.8	0.0	0.0	0.0	0.0	41.9	0.1	0.0	0.0	22.5	0.5
MICROSCOPE	0.0	0.0	0.0	0.0	0.0	0.0	0.0	0.0	0.0	0.0	0.0	0.0	0.2	0.1

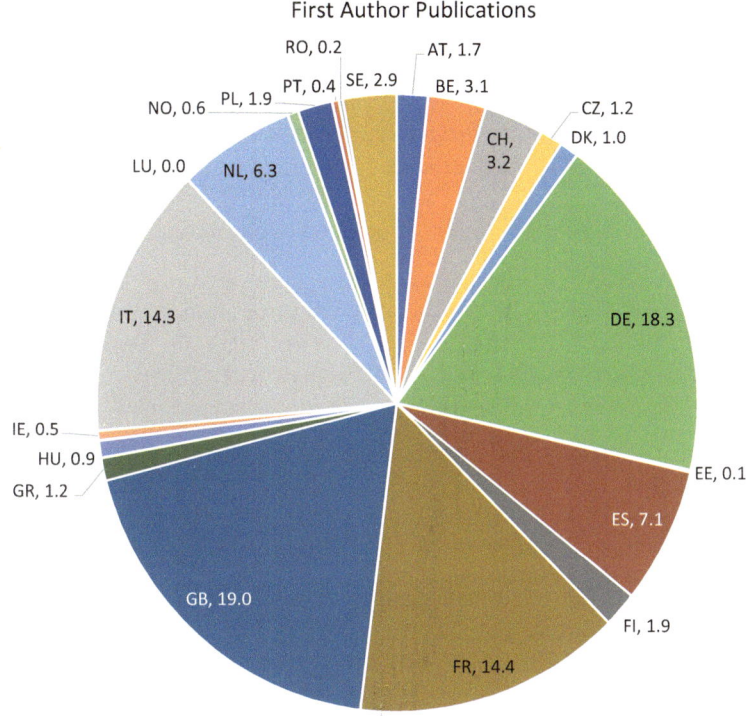

Fig. 2.3 Percentages of first author publications for the ESA-led missions for each ESA Member State

used. These are indicated as "% Contrib" in Table 2.3. Since the contributions are proportional to the Gross Domestic Product (GDP) of each Member State, they give an indication of the relative economic weights of the Member States. Dividing the relative publication numbers by the financial contributions gives a measure of how a Member State's scientists were able to exploit and lead (as first authors) the opportunities afforded by the Science Programme compared to the financial contribution that their country makes to the Programme.

Figure 2.4 shows, for each Member State, the relative fraction of first-author publications as a function of the fraction of their contribution to the ESA Science Programme. The distribution appears to follow a linear correlation in agreement with the straight line, which shows equal contributions to publications and finances, although some scatter is clearly present. This is easier to see in Fig. 2.5, where we take the ratio of the two quantities shown in Fig. 2.4 to better highlight the scatter as well as the uncertainties (which, as already mentioned, are assumed to follow a Poisson distribution on the number of publications).

Examining the four countries that each contributed more than 10% of the Science Programme's finances reveals that scientists located in the United Kingdom,

Germany, France, and Italy had 4362, 4195, 3293 and 3275 first author publications, respectively (see Table 2.3). Comparing these numbers with the financial contributions to the Science Programme (as we do in Figs. 2.4 and 2.5) shows that scientists located in the United Kingdom and Italy both produced 23% more first author publications than predicted by these countries contributions. Similarly, scientists located in France and Germany produced 4% fewer and 14% fewer first author papers than predicted by their contributions, respectively.

Of the other ESA Member States, three stand out as having exceptionally high publication-to-contribution ratios:

1. The Netherlands has 1439 first author publications which is 42% more than predicted.
2. Finland has 428 publications which is 40% more than predicted.
3. Hungary has 206 publications which is 45% more than predicted.

We note that of the 206 first author publications from Hungary a high fraction (56 publications) use results from Gaia and 31 use Herschel results. It is noticeable that some of the smaller Member States have low first author publication-to-contribution ratios (see Fig. 2.5):

1. Denmark with 229 first author publications which is 43 % less than predicted.
2. Norway with 131 first author publications which is 74 % less than predicted.

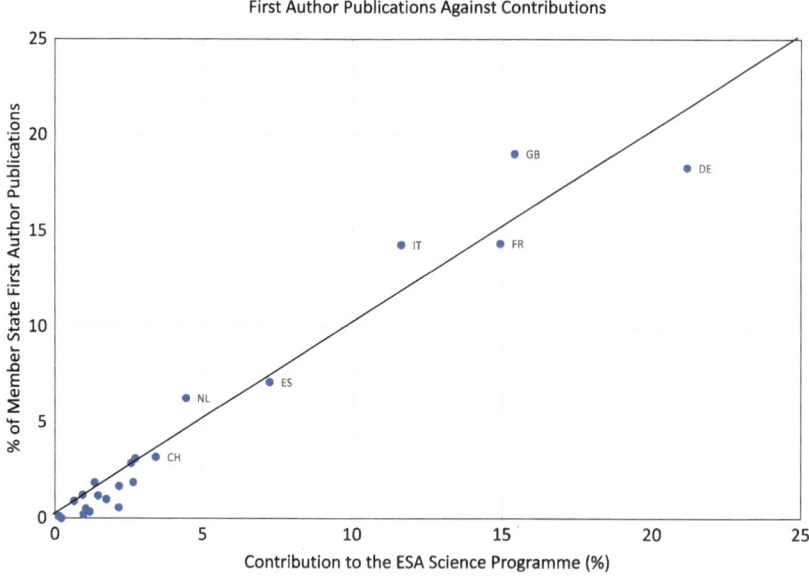

Fig. 2.4 Percentages of first author publications for each ESA Member State as a function of the 2000–2021 financial contributions of those countries to the Science Programme. The seven countries making the largest financial contributions to the ESA Science Programme are labelled . Error bars are too small to be visible

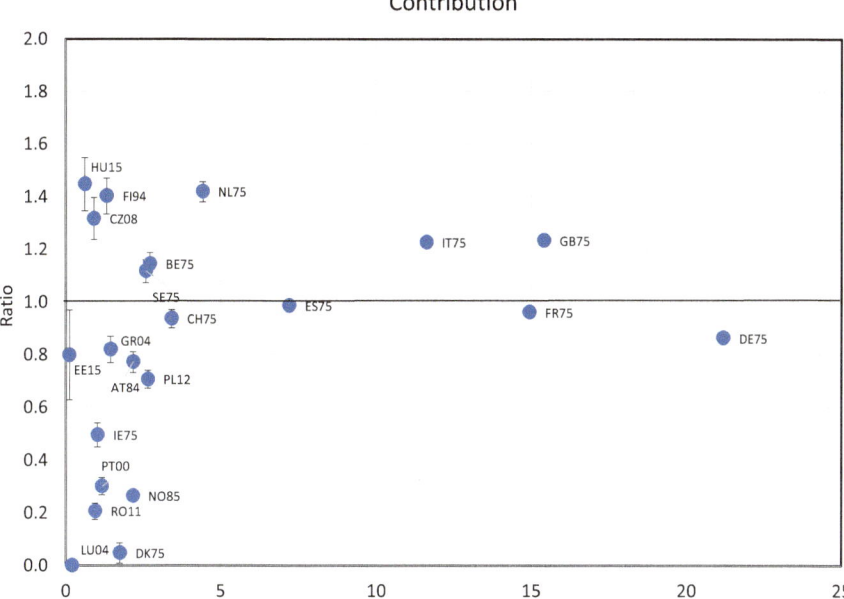

Fig. 2.5 Ratio between the number of first authors located in each ESA Member State and the average 2000–2021 financial contributions of those countries to the ESA Science Programme. Error bars are only to "guide the eye" as they assume that the publication numbers follow a Poisson distribution. Next to the ISO 3166 country codes of each Member State, we also indicate the last two digits of the year in which they joined ESA

3. Ireland with 117 first author publications which is 50 % less than predicted.
4. Portugal with 81 first authors which is 70 % less than expected.

We have also examined the publication statistics using all authors normalised so that each publication contributes 1.0 authors (Tables 2.4 and 2.5). This is our preferred method of measuring the *overall* contribution of a country whereas the number of first authors provides a better measure of "leadership" in data exploitation. The above conclusions are not changed significantly for the four largest Member States if this metric is used.

Indeed, the normalised fraction of papers per Member State divided by their contribution to ESA shown in Fig. 2.6 reveals that the large scatter remains, particularly for smaller Member States. The largest ratio changes are seen from Hungary where the ratio decreases from 1.45 for first author publications to 1.19 when all authors are considered and for the Czech Republic where this ratio decreases from 1.31 to 1.18 (see also Tables 2.3 and 2.5).

Considering first-author papers, most of the founding Member States (all of which joined ESA when it was created in 1975) have ratios within ~20% of one.

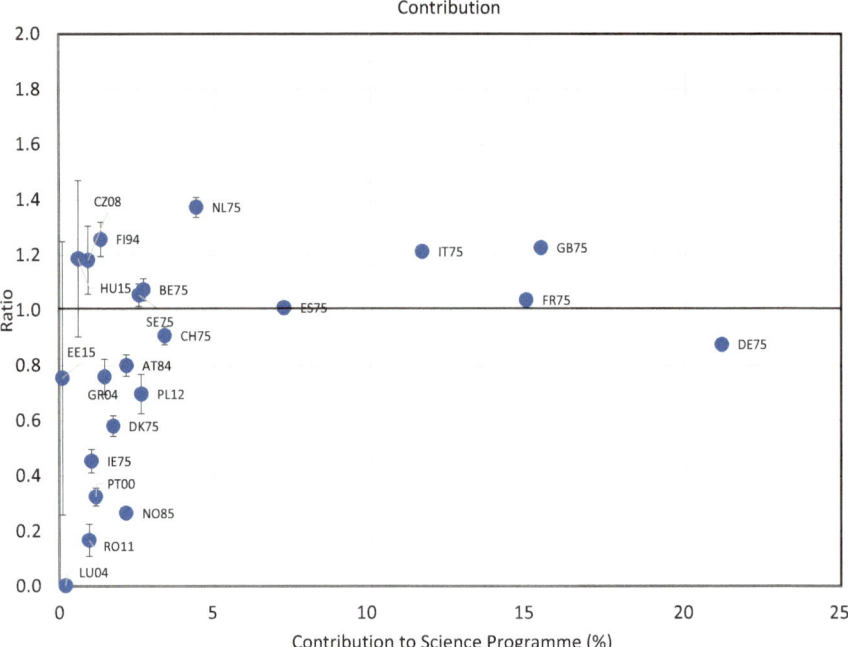

Fig. 2.6 Ratio between the number of all authors located in each ESA Member State and the average 2000–2021 financial contributions of those countries to the ESA Science Programme. Error bars are only to "guide the eye" as they assume that the publication numbers follow a Poisson distribution. Next to the ISO 3166 country codes of each Member State, we also indicate the last two digits of the year in which they joined ESA

Larger variations are seen for two of the founding Member States, namely Denmark and Ireland, with ratios of ∼0.05 and ∼0.5, respectively. This might appear unusual if one were to assume that, after almost 40 years, all founding Member States should have reached a similar level of engagement by their communities in space science research and technology. However, it is also clear that for smaller Member States there is a very large scatter and there does not appear to be a direct correlation between the year in which a Member State joined and the ratio of first- or all-authors in publications. To clarify this, we have added to the country codes in Figs. 2.5 and 2.6 the last two digits of the year in which they joined ESA. The lack of a correlation between ratio of first-authors and "space maturity" can be seen for instance by comparing Hungary, the Czech Republic, and Finland, which joined, respectively, in 2015, 2008, and 1994: these three countries have some of the highest rates, comparable to that of the Neherlands, which as a founding member joined ESA in 1975.

Therefore, the evaluations of the number of expected papers in the analysis above, particularly for smaller Member States, are at best an indication, because they are normalised by the GDP of each country. To be sure, a number of additional factors besides the GDP can affect the productivity of a scientific community in a particular country and field of research. These could include for instance the scientific and technical priorities of the country, the size and specialisation of its scientific communities, as well as the historical heritage. Therefore, the statistics provided here are not meant to be used to assess or judge the scientific performance of the space science communities in the ESA Member States. Instead, they might serve as a long term tool to help identify potential untapped areas of space research and technology development. Additionally they can also serve as a reference for future studies.

2.4 Publication Metrics

One of the advantages of using ADS is that it provides a number of powerful metrics that can be easily used to provide insights into the publications in a library [4]. For each of the ESA- and partner-led missions, the following were extracted from the ADS:

1. The numbers of refereed publications with >100 citations (i100). This is often used as a measure of the number of highly influential publications.
2. The number of citations (from refereed and non-refereed publications) to the publications in the libraries
3. The Hirsch [12] or h-index, which is the position in a citation ranked list where the rank equals the number of citations. For example a facility or individual with an h-index of 100 would have 100 publications each with 100 or more citations. This index gives an estimate of the importance and broad impact of a mission's cumulative results.
4. The m-index. Since the h-index can only increase with time, the m-index, defined as the h-index divided by the time between the first and last publications of a mission, is also used to provide an estimate of the productivity of a mission.
5. The 2-year impact factor for each mission with sufficient publications in 2019 and 2020. The impact factor used is the number of citations in 2021 to the mission's refereed publications in 2019 and 2020, divided by the number of such publications in ADS. It may be regarded as a measure of the importance of the scientific output of a mission in the years immediately following publication.

In addition, a dedicated software tool was used to calculate:

1. The percentage of publications whose first authors are located in institutes in the ESA Member States, compared to the total.
2. The number of unique author names from the publications for each mission. This is not the same as the number of first authors as e.g., A.N. Other and A. Other will be counted as two authors, while multiple authors who have the same name will be under-counted.

Table 2.12 Publication metrics for the ESA-led missions arranged in order of launch date. For missions still in operation, only the launch date is given under "Operations". The total number of refereed publications is given since launch. The number of citations is the number of citations from both refereed and non-refereed publications to the mission's publications. MS/Total is the number of first authors from institutes located in the ESA Member States divided by the number of first authors from all countries. The 2-year impact factor is the number of citations in 2021 to the publications from a mission in 2019 and 2020 divided by the number of such publications

ESA-led Mission	Operations	Refereed Publications			Authors	Citations	Indicators		
		Total No.	No. with >100 citations	MS/Total First Authors	No. Unique Names		h-index	m-index	2-year Impact Factor
COS-B	1975-1982	146	12	81%	380	5872	12	0.8	-
IUE	1978-1996	5105	399	39%	8355	204250	169	3.9	4.2
Exosat	1983-1986	743	52	76%	1122	26770	79	2.2	2.0
Giotto	1985-1992	266	40	71%	513	14988	62	2.1	-
Hipparcos	1989-1993	2692	264	46%	6724	121977	153	4.8	6.6
Ulysses	1990-2005	1950	118	38%	3302	61066	107	3.7	6.8
ISO	1995-1998	2062	218	60%	5516	106466	136	5.4	3.9
SOHO	1995-	6252	473	35%	8905	229472	172	6.6	4.0
Huygens	1997-2005	222	12	52%	802	7772	46	2.2	1.0
XMM-Newton	1999-	6963	500	55%	18648	261004	178	8.5	6.3
Cluster	2000-	2962	132	40%	4736	76682	109	4.4	4.1
INTEGRAL	2002-	1904	145	61%	11213	78067	115	6.1	6.9
SMART-1	2003-2006	70	0	70%	352	795	17	1.0	0.3
Mars Express	2003-	1464	70	54%	3793	44729	88	4.9	3.6
Rosetta	2004-2016	1314	48	67%	3831	29049	77	4.5	3.5
Venus Express	2005-2014	737	15	53%	2135	15797	55	3.7	3.2
Herschel	2009-2013	3284	235	58%	11801	125619	137	12.5	5.1
Planck	2009-2013	2742	278	43%	9424	158333	155	15.5	17.4
PROBA-2	2009-	137	2	59%	747	2644	28	2.8	3.2
Gaia	2013-	6403	178	41%	25107	144049	123	17.6	8.1
LISA Pathfinder	2015-2017	33	2	81%	275	753	9	1.8	1.8
ExoMars 2016	2016-	138	0	56%	1032	1432	20	2.9	7.8
BepiColombo	2018-	71	0	76%	947	419	10	3.3	3.4
CHEOPS	2019-	38	0	95%	448	266	9	9.0	-
Solar Orbiter	2020-	114	1	72%	1684	1206	17	17.0	-

Table 2.12 shows the metrics for the ESA-led missions. It can be seen that the largest numbers of highly influential publications (i100) are from XMM-Newton [6] and SOHO [7] with 500 and 478, respectively. Of the older missions, IUE [13] has 399 such publications and Hipparcos [10] 264. Publications from the more recently launched Herschel [14] and Planck [15] missions (both launched in 2009) and Gaia [8] (launched in 2013), have had less time to accumulate citations, so their values of i100 of 235, 278 and 178, respectively are also very impressive.

There are large variations in the percentage of ESA Member-State first authors compared to the total first authors for the different missions with the average Member state/total first authors of 59% for the ESA-led missions. This illustrates the global nature of the exploitation of the results from ESA's Science Programme missions. As expected, missions that are major collaborations with other agencies, such as IUE [13], SOHO [7], and Ulysses [16] have lower ESA Member State first author percentages of 39, 35, and 38%, respectively. In terms of absolute numbers of publications, from the ESA-led missions that are not in collaboration, then Gaia [8] and XMM-Newton [6] have provided 3778 and 3133 non-ESA Member State first author publications through to 2021.

The number of unique author names is a measure of the size of the community exploiting data from a mission. The numbers given in Table 2.12 need to be interpreted with caution. An author with publications under A. Other and A.N. Other will appear twice, while authors with the same name will be only counted once. To obtain an estimate of how big this effect is, the 1120 unique names for Exosat [17] were found to correspond to 929 authors, which is 21% higher than the number of unique names. Thus, this metric is best used for comparison between missions, rather than an absolute measure of community size. Indeed, missions such as INTEGRAL [18] have benefited from contributing to several multi-messenger publications with ~1000 authors each, and it could be argued that this does not reliably reflect the users of the mission itself. Unsurprisingly, Gaia [8] and XMM-Newton [6] have the largest numbers of unique author names (25,107 and 18,648), respectively.

The long-lived XMM-Newton [6] and SOHO [7] missions dominate the citation statistics with 261,004 and 229,472, citations respectively. There are also strong showings from IUE [13], Hipparcos [10], Herschel [14], Planck [15] and Gaia [8].

For the three indicators, the h-indices are dominated by the long-lived missions XMM-Newton [6], SOHO [7], and IUE [13], but the more recent Herschel [14], Planck [15], and Gaia [8] missions do well explaining their high m values. The mission with the highest 2-year impact factor is Planck [15] (17.4) followed by Gaia [8] and ExoMars 2016 [19], with impact factors of 8.1 and 7.8, respectively. The impact factors of Hipparcos [10] (1989–1993) and Ulysses [16] (1990–2005) remain an impressive 6.6 and 6.8, respectively. The INTEGRAL [18] impact factor of 6.9 is in part due to a number of multi-messenger publications to which the mission contributed.

Table 2.13 shows the same metrics as above for the partner-led missions. HST dominates the statistics with 2474 publications with more than 100 citations (i100) followed by Cassini [9] with 133. If the missions that are led by an agency from an ESA Member State are ignored (CoRoT [20] and MICROSCOPE [21]) then 29 % of the first authors from the remaining missions are located in the ESA Member States. In terms of absolute numbers, the 34 % of HST first authors located in the ESA Member States means a total of 6573 publications led by authors from ESA Member State institutes through to 2021. Cassini [9] has provided 757 such publications and Hinode [22] 519, also illustrating their importance to ESA Member States scientists. HST also dominates the citation numbers, with more than a million citations to the publications from the mission and it has the highest h-index of any of the missions (324).

Table 2.13 Publication metrics for the partner-led missions arranged in order of launch date. For missions still being operated, only the launch date is given under "Operations". The total number of refereed publications is given since launch. The number of citations is the number of citations from both refereed and non-refereed publications to the mission's publications. MS/Total is the number of first authors from institutes located in the ESA Member States divided by the number of first authors from all countries. The 2-year impact factor is the number of citations in 2021 to the publications from a mission in 2019 and 2020 divided by the number of such publications

Partner-Led Mission	Operations	Refereed Publications			Authors	Citations	Indicators		
		Total No.	No. with >100 citations	MS/Total First Authors	No. Unique Names		h-index	m-index	2-year Impact Factor
HST	1990-	19333	2474	34%	38800	1057706	324	10.8	9.6
Cassini	1997-2017	2227	133	34%	4027	79268	111	4.6	2.9
Double Star	2003-2007	161	3	39%	628	2835	30	1.8	4.7
Suzaku	2005-2015	1139	54	19%	4053	35458	81	5.4	4.0
AKARI	2006-2011	1086	32	30%	5320	25888	70	4.7	4.3
Hinode	2006-	1442	75	36%	2441	51382	92	6.6	3.6
CoRoT	2006-2013	238	20	78%	995	10477	55	3.9	3.7
Chandrayaan-1	2008-2009	164	1	24%	723	3186	30	2.3	2.4
IRIS	2013-	449	8	35%	1007	10719	49	6.1	5.4
Hitomi	2016-2016	74	2	13%	821	1109	14	2.8	2.4
MICROSCOPE	2016-2018	17	1	94%	71	491	9	1.8	7.0

Figures 2.7, 2.8, 2.9, 2.10, 2.11, 2.12, 2.13, 2.14, 2.15, 2.16, 2.17, 2.18, 2.19, 2.20, 2.21, 2.22, 2.23, 2.24, 2.25, and 2.26 show the evolution of selected metrics against year after launch for ESA-led missions. The time when each mission was operational is also shown in the upper panels. For each mission, as well as the annual number of refereed publications, the number of citations per year, the annual number of i100 publications and the 2-year impact factor (the number of citations in year n to the publications in years $n-1$ and $n-2$ divided by the number of such publications) are shown when there are sufficient statistics to evaluate these parameters.

These figures reveal a number of interesting insights:

1. For many missions the number of publications per year increases rapidly after the first $\sim$2–3 years and remains high during operations before decreasing slowly after the end of operations. This "long tail" of publications can last for many years, e.g., 25 years after the end of operations there are still $\sim$25 refereed publications per year from IUE [13], some 20% of the peak publication rate (Fig. 2.8). This emphasises the importance of long-term scientific archives that provide continued access to relevant data.

2. In contrast, the publication rates from Hipparcos [10], Planck [15] and Gaia [8] took longer to increase towards their peaks. This is likely due to it being necessary to accumulate and analyse all-sky data sets before detailed scientific analysis could be performed (see Figs. 2.11, 2.25, and 2.26).

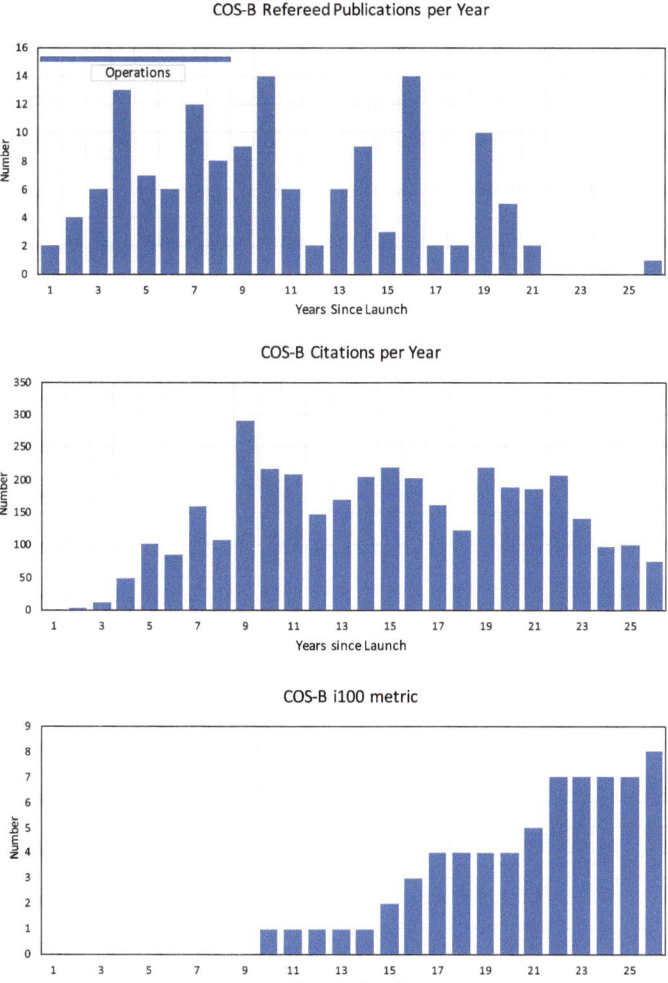

Fig. 2.7 COS-B publication metrics

3. For the first 11 years of the Rosetta [23] mission the publication rate remained relatively low (Fig. 2.22) during and after the encounters with the Steins and Lutetia asteroids. This changed once the mission encountered its main target comet 67P/Churyumov-Gerasimenko, which provided for a marked increase in the publication rate.
4. The scientific importance of a number of missions can be seen in the very high number of citations many years after operations started. In particular, the citation rates for four missions still in operations have grown to impressive numbers: XMM-Newton [6] has around 25,000 citations per year after 22 years of operations, SOHO [7] has around 14,000 citations per year after 26 years

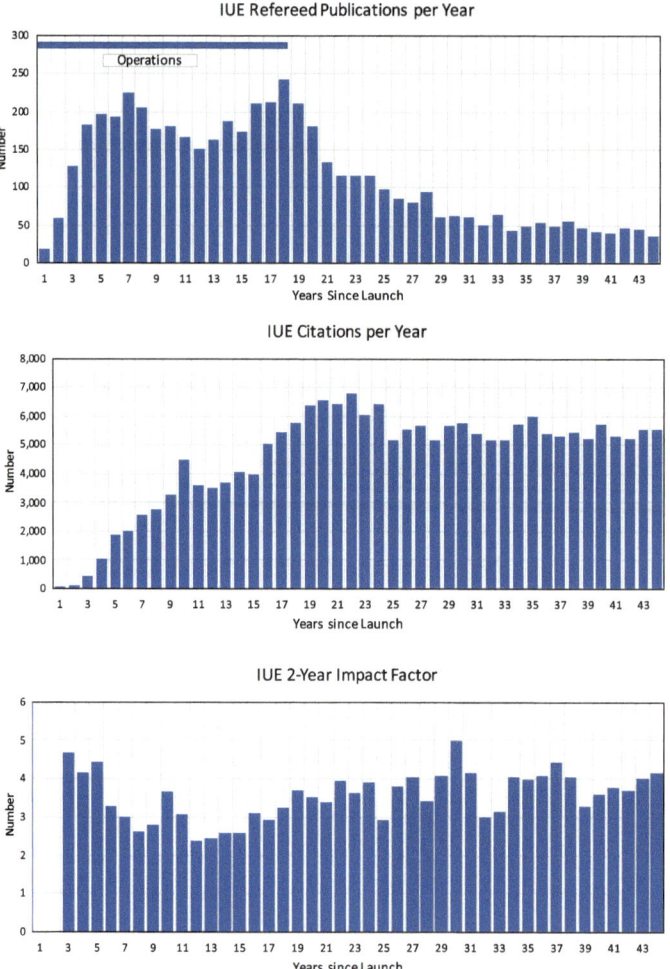

Fig. 2.8 IUE publication metrics

of operations, Cluster [11] has around 8000 citation per year after 22 years of operations as does INTEGRAL [18] after 19 years of operations.
5. The longevity of the scientific relevance in IUE [13], Hipparcos [10] and Infrared Space Observatory (ISO) [24] results is remarkable; IUE [13] and Hipparcos [10] continue to have around 5000 citations per year some 25 years after operations ended. Similarly, there are around 4000 citations per year to the ISO [24] publications some 26 years after operations ended. Both Planck [15] and Herschel [14] have impressive citation statistics having some 30,000 and 15,000 citations per year some nine years after the ends of operations.

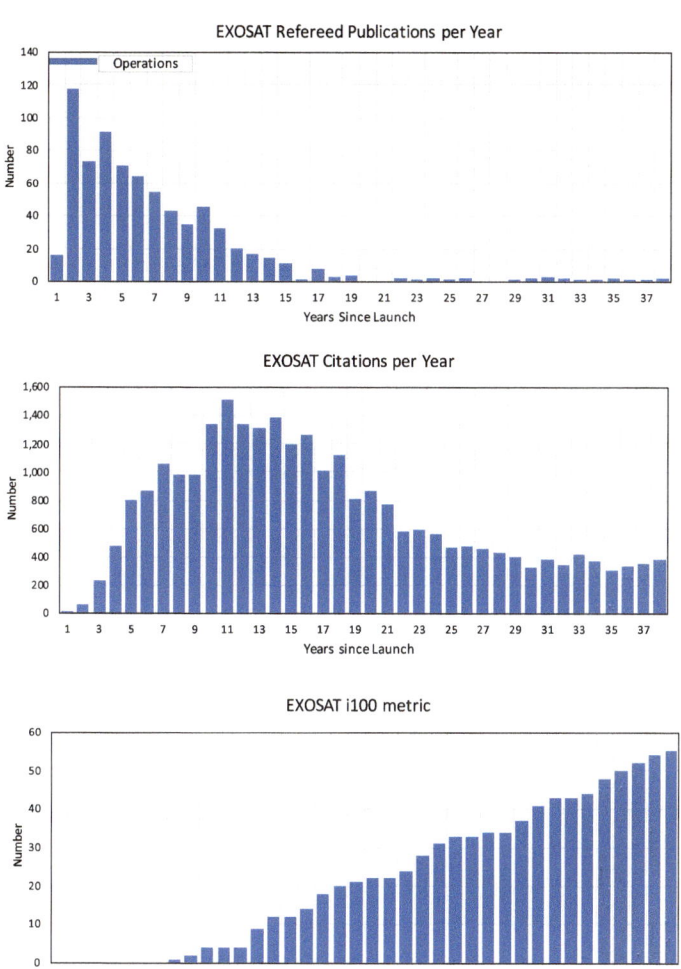

Fig. 2.9 Exosat publication metrics

It is interesting to compare how i100, the number of citations, the h-indices and the 2-year impact factor evolve with time for the ESA-led missions. These were extracted from ADS and are shown in Figs. 2.27, 2.28, 2.29, and 2.30. In order to compare missions, the values are shown against years since launch up to a maximum of 20 years. Solar Orbiter is not included as it was launched too recently to have meaningful data.

Figure 2.27 shows how the number of publications with >100 citations (i100) evolves with time after launch for selected ESA-led missions. It can be seen that this metric increased most rapidly for Gaia [8], Planck [15] and Herschel [14] followed by XMM-Newton [6] and SOHO [7] with XMM-Newton having the

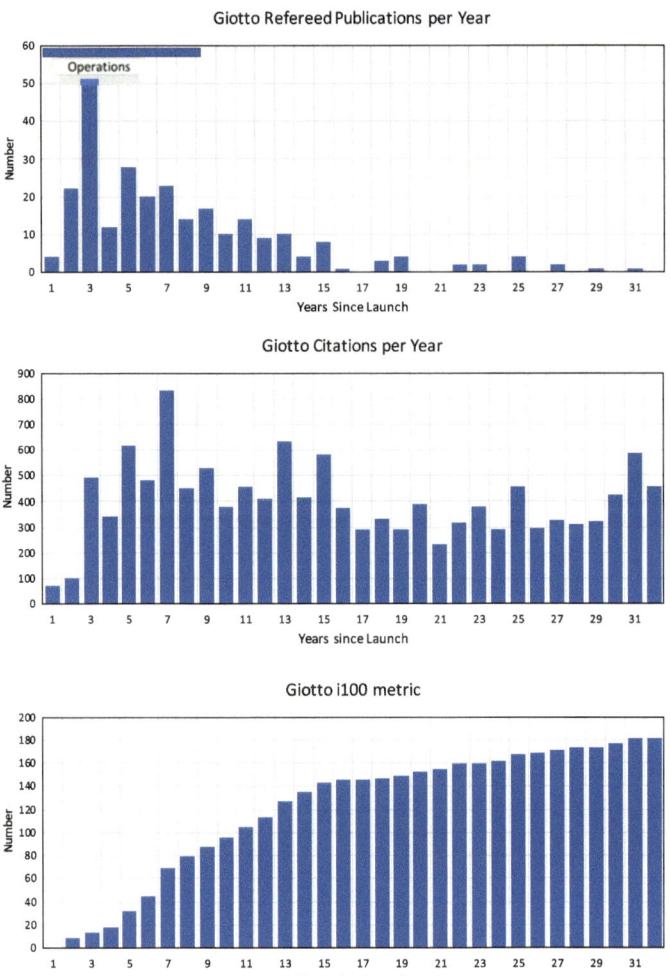

Fig. 2.10 Giotto publication metrics

highest absolute value. The evolution of this metric illustrates the high interest in the results from these missions as well as illustrating that it takes typically three to four years from launch before significant numbers of such publications start to appear.

Figure 2.28 shows how the number of citations for selected ESA-led missions evolved with time. The rapid increase in the number of citations from Gaia [8] is evident as are the contributions of Planck [15], Herschel [14], XMM-Newton [6] and SOHO [7]. Unsurprisingly these same missions had also the highest numbers of i100 publications (see Fig. 2.27).

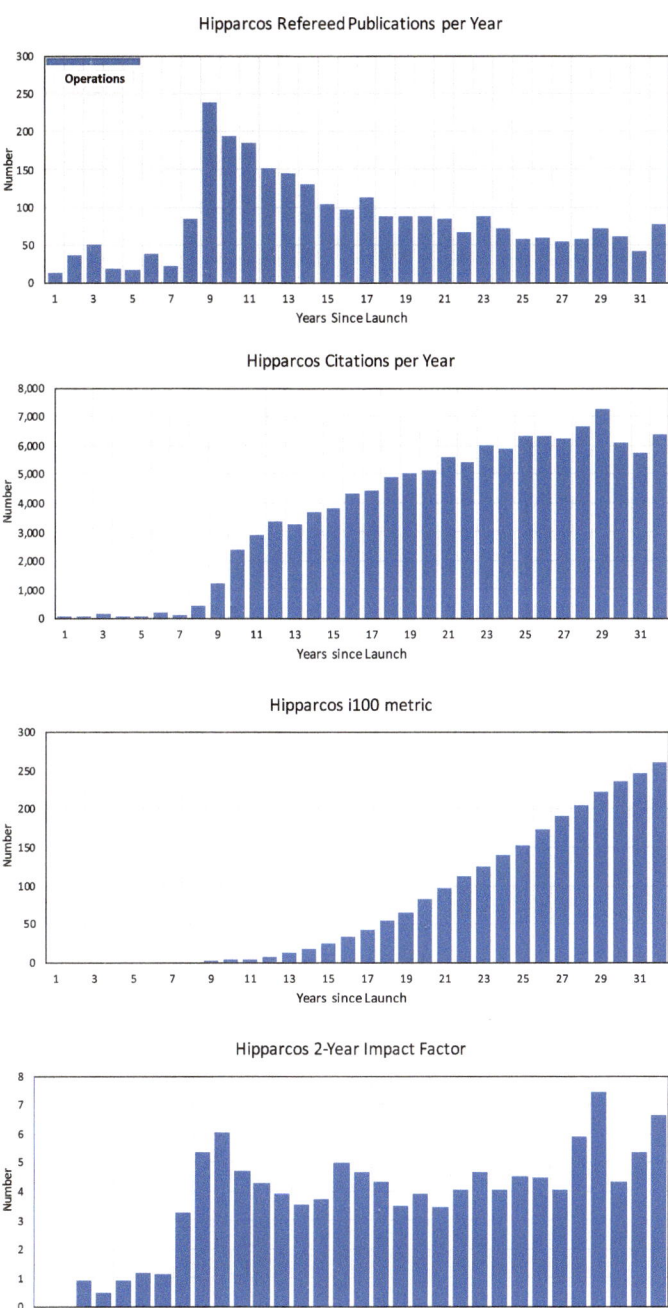

Fig. 2.11 Hipparcos publication metrics

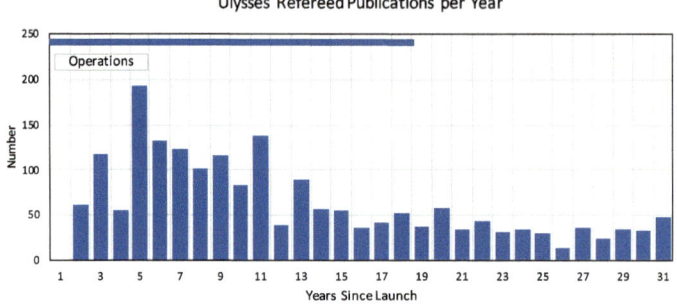

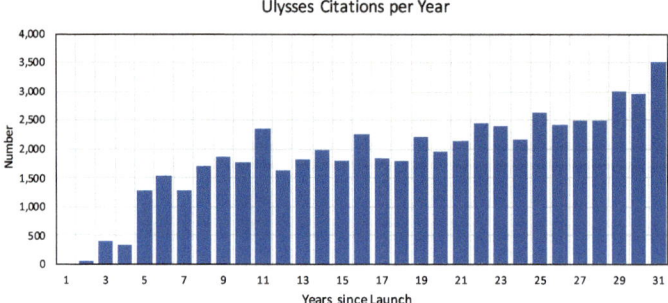

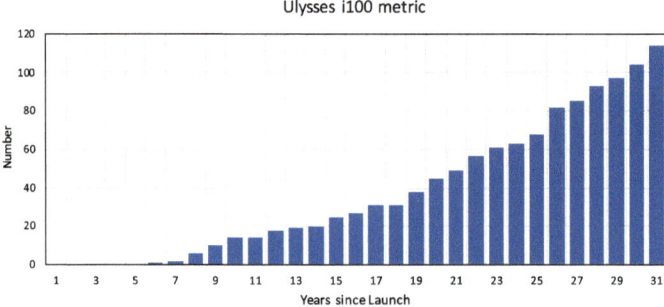

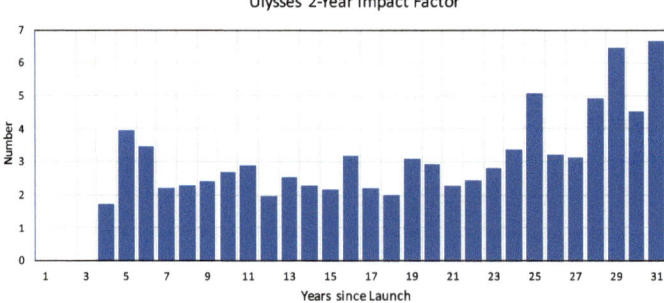

Fig. 2.12 Ulysses publication metrics

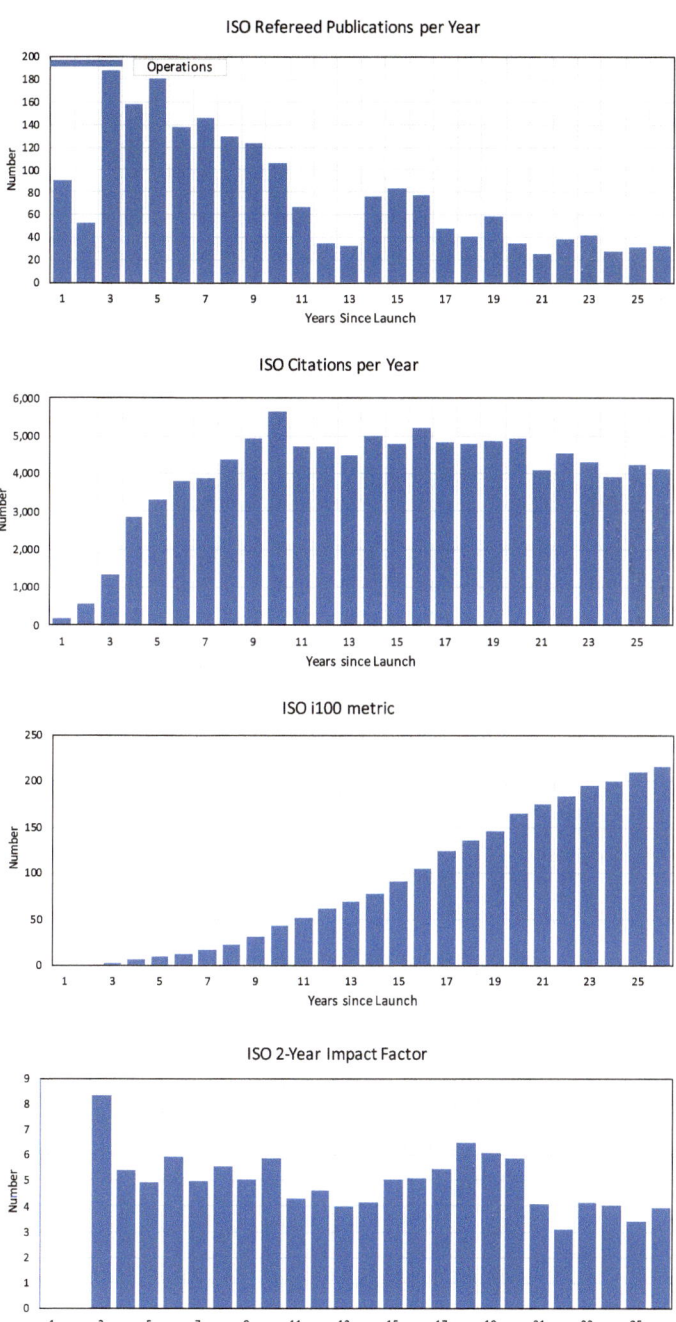

Fig. 2.13 ISO publication metrics

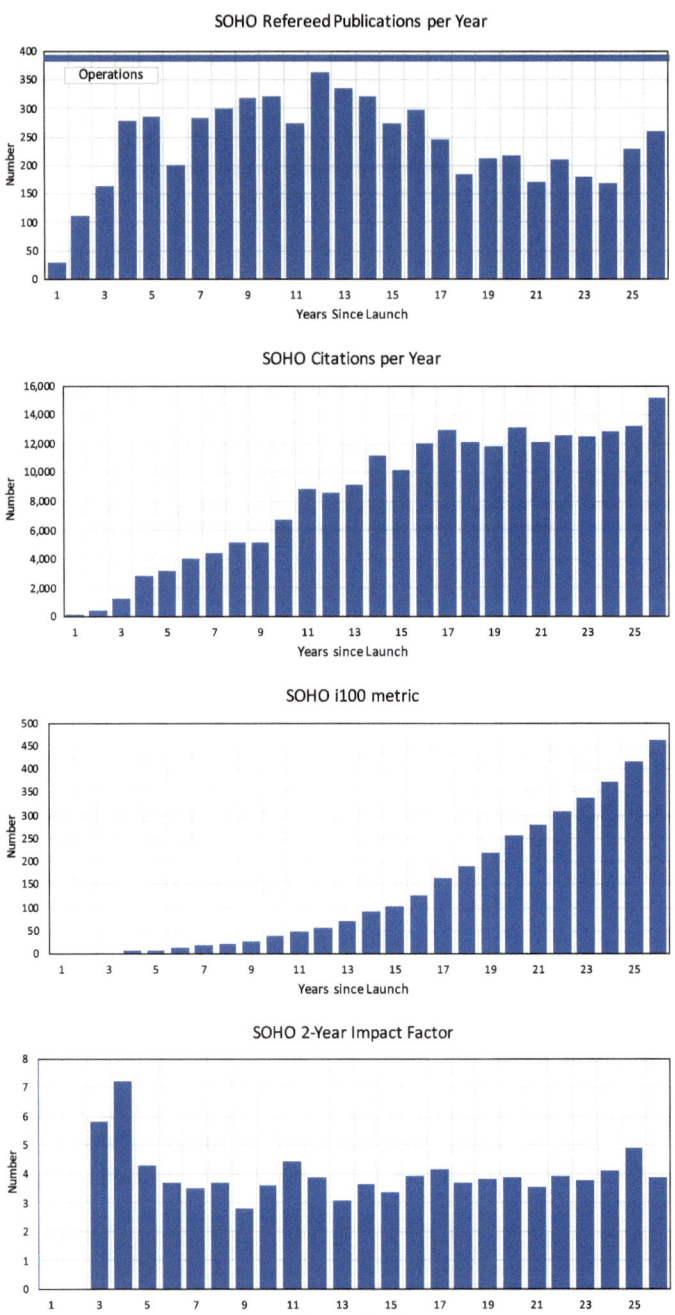

Fig. 2.14 SOHO publication metrics

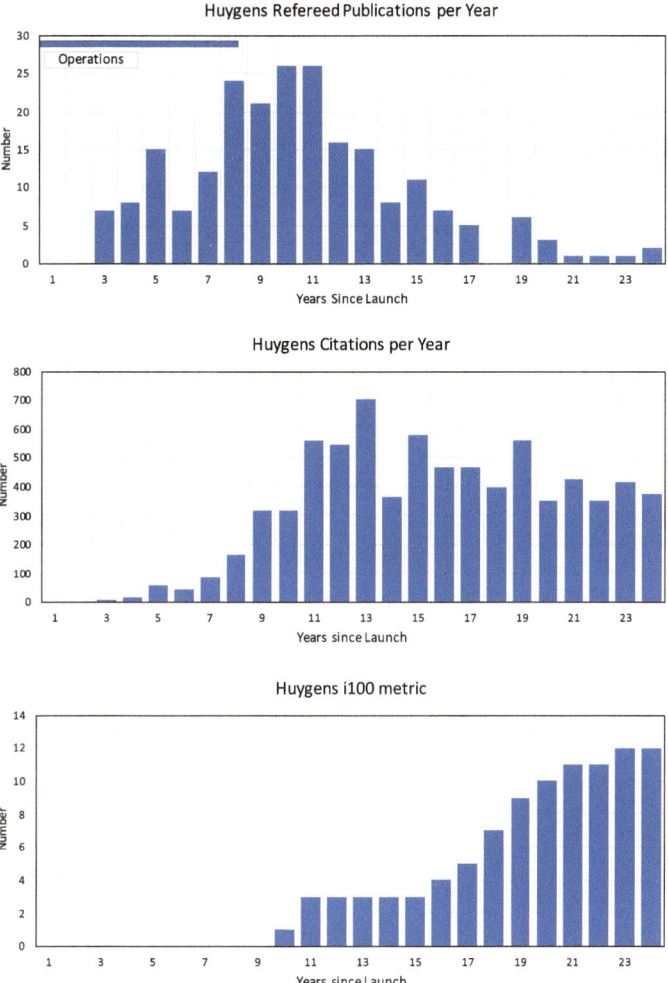

Fig. 2.15 Huygens publication metrics

Figure 2.29 shows the evolution of the h-indices with time for selected ESA-led missions. The rapid increase in h-indices for Gaia [8], Planck [15] and Herschel [14] is evident, as are the high absolute values for XMM-Newton [6] and SOHO [7].

Figure 2.30 shows how the 2-year impact factors evolve with time for a selection of ESA-led missions. The impact factors typically rise quickly to maxima between 3 to 5 years after launch and then slowly decay. The effect on the impact factors of the cometary encounter for Rosetta [23] and the advent of multi-messenger astronomy for INTEGRAL [18] are clearly evident in this figure.

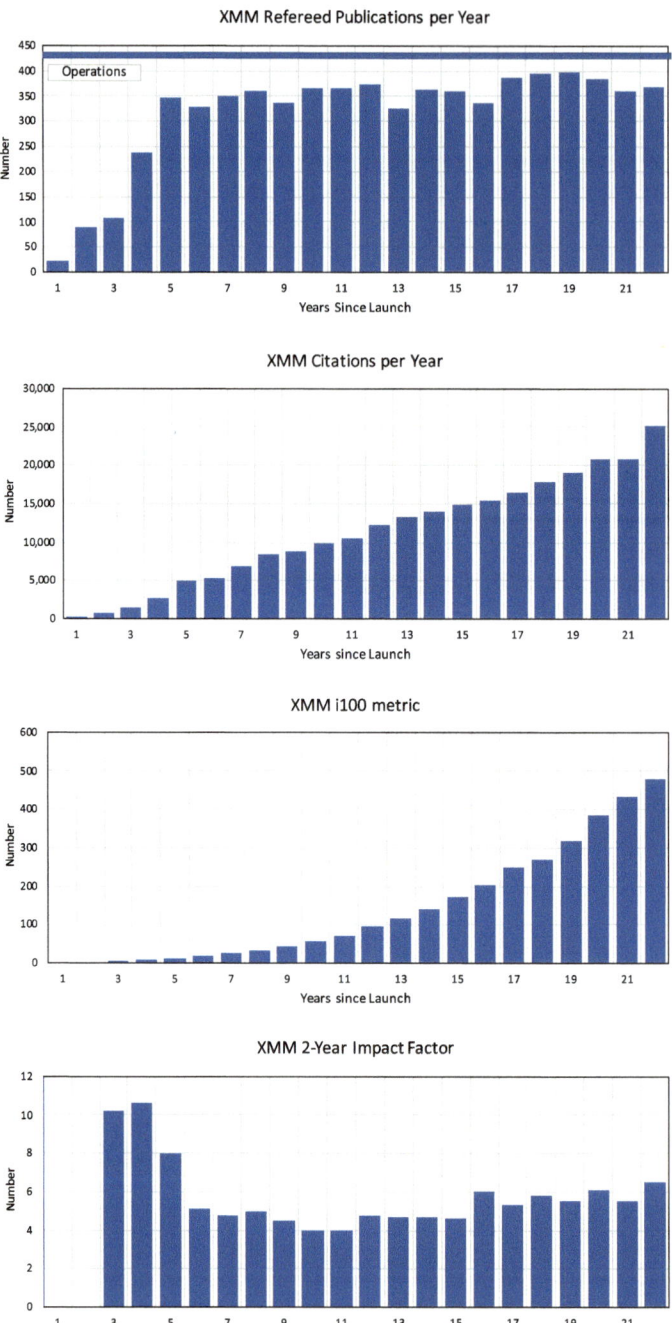

Fig. 2.16 XMM-newton publication metrics

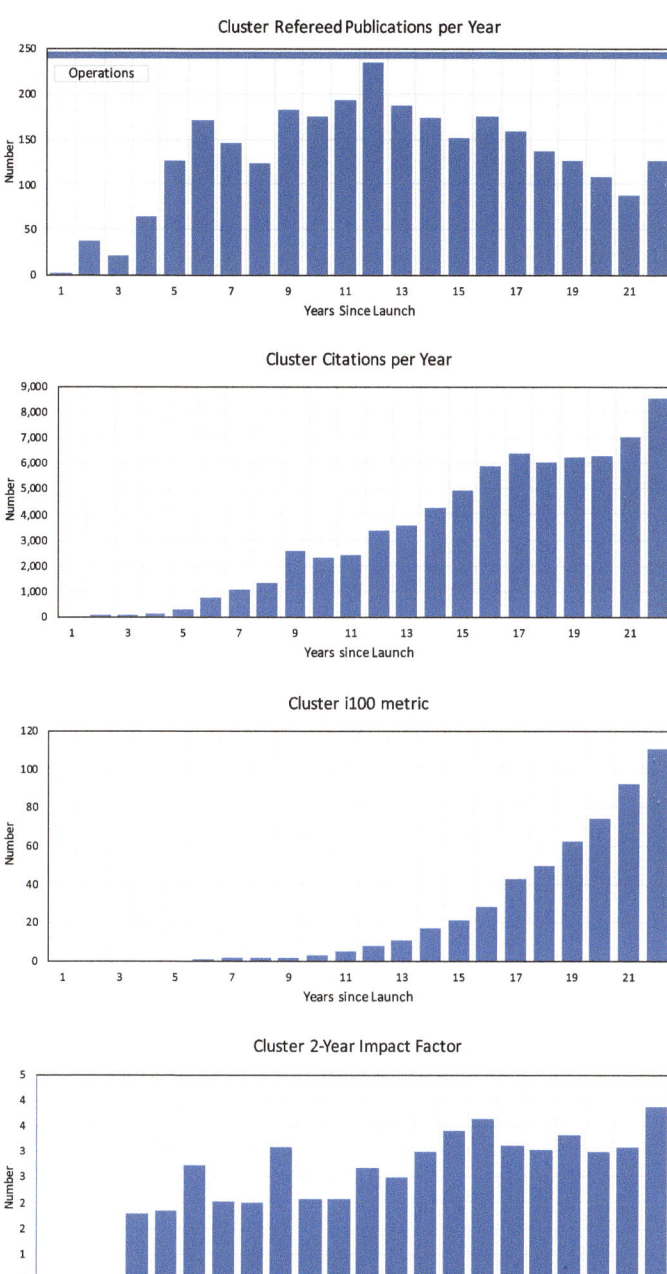

Fig. 2.17 Cluster publication metrics

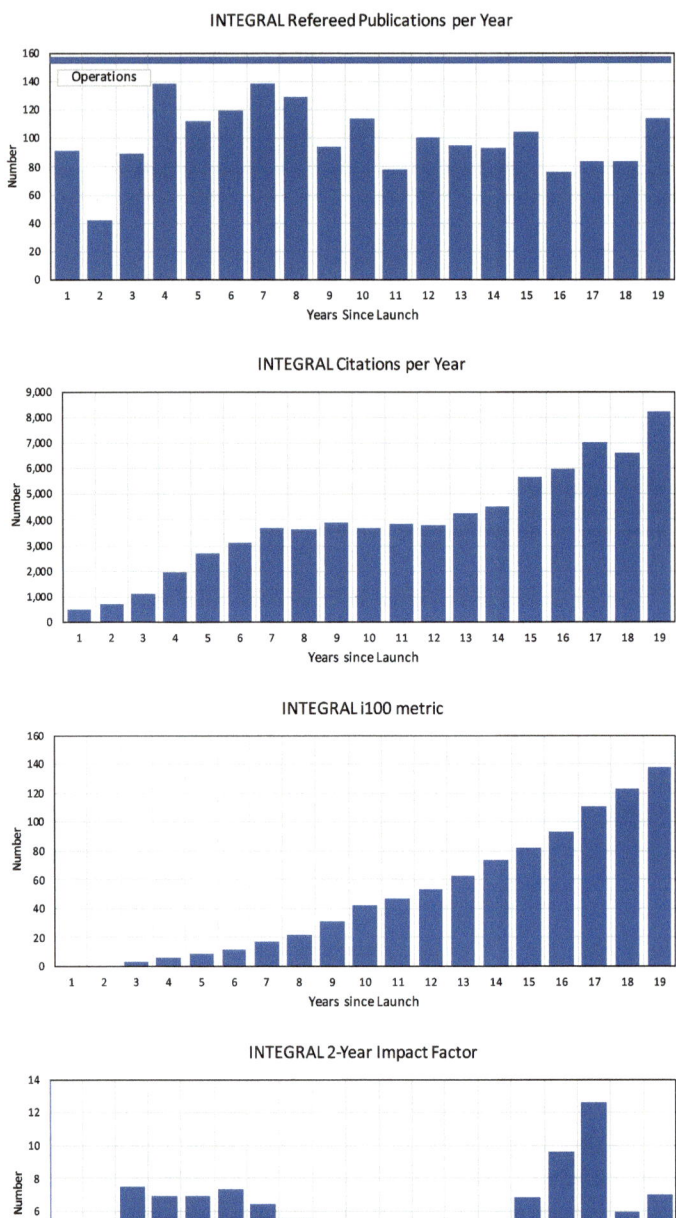

Fig. 2.18 INTEGRAL publication metrics

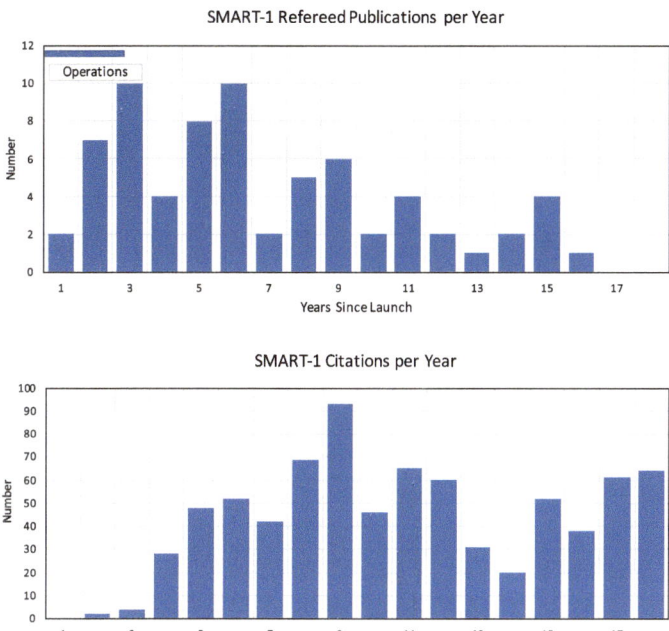

Fig. 2.19 SMART-1 publication metrics

2.5   Cross Mission Publications

The publications libraries are built per mission, using uniform search criteria (see Sect. 2.2), so publications making use of data from more than one mission can be identified by simply cross-matching the individual libraries. In Fig. 2.31 we show the results of this analysis in graphical form for a sample of astronomy and solar system missions, both ESA-led and partner-led. In each panel, a specific colour is assigned to each mission. The total number of papers for each mission, in a five year period between 2016 and 2021, is indicated at the top of the corresponding mission's column. The fraction of papers in common with other missions is indicated by the height of the other colour bars in each mission column. For instance, of the 4620 Gaia papers in that period, about 7.6% also used data from HST, 2.6% from XMM-Newton, 1.5% from Herschel, 1.0% from Planck, and 0.3% from ISO. For the astronomy missions in Fig. 2.31, the average fraction of cross-mission papers, weighted by the number of papers, is about 18%, but can be considerably larger for legacy missions, namely 28 % for Herschel and over 50% for ISO. One possible interpretation is that one should expect more overlap between missions that cover similar wavelengths ranges (e.g., ISO and Herschel), but it is also possible that the overlap with legacy missions is larger because thanks to the ESA Space Science Archives their data remain available and are used as a reference or comparison

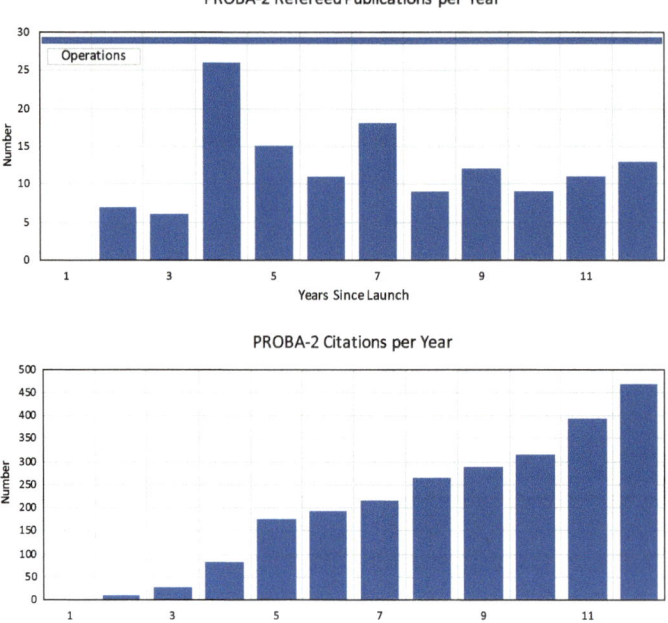

Fig. 2.20 Proba-2 publication metrics

stone when data from new missions become available in a similar or complementary
wavelength range. The fraction of papers in common for Solar System missions
is somewhat lower, on average 9% for heliophysics and 5% for planetary science
missions, but again it can be very large for missions covering the same or similar
fields (e.g. ExoMars and MEX, or Hinode and SOHO). In general, this underlines
the importance of preserving data from legacy missions, which can enrich the
scientific interpretation of more recently acquired data.

2.6   Archival Publications

The ESA Space Science Archives serve as the single access point for all scientists
interested in observations and measurements collected by current and legacy ESA
Science Programme missions. When scientists who are not involved in the original
research projects or in the provision of on-board instruments retrieve those data and
use them for new investigations, they help increase the value of the missions. They
contribute to the advancement of science through independent analysis of existing
data to answer a range of scientific questions. In this way, investigations based
on archival data effectively augment the scientific return of the ESA missions. We

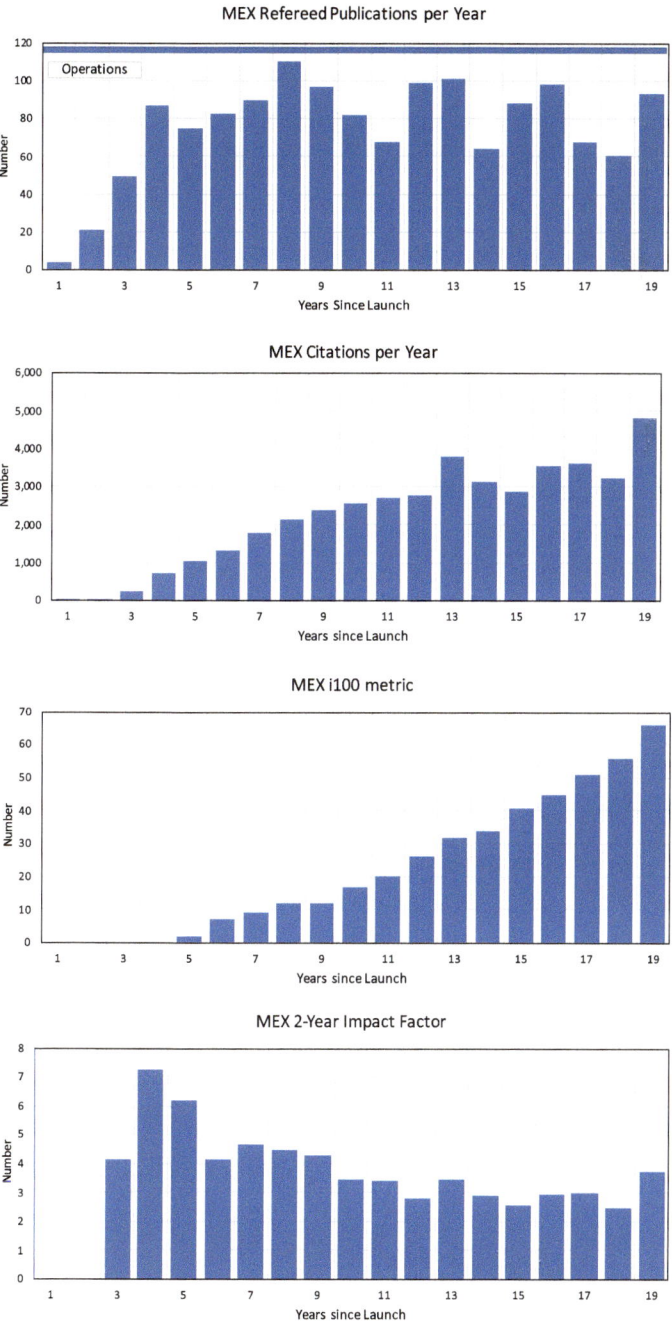

Fig. 2.21 MEX publication metrics

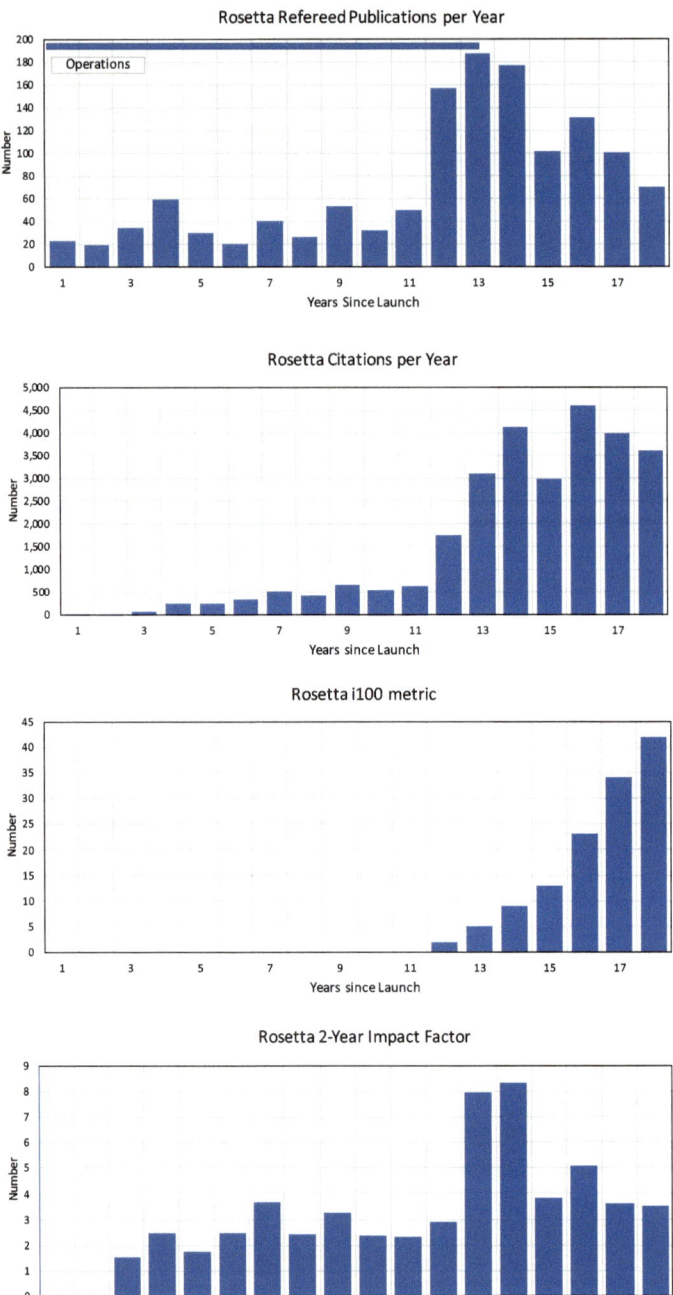

Fig. 2.22 Rosetta publication metrics

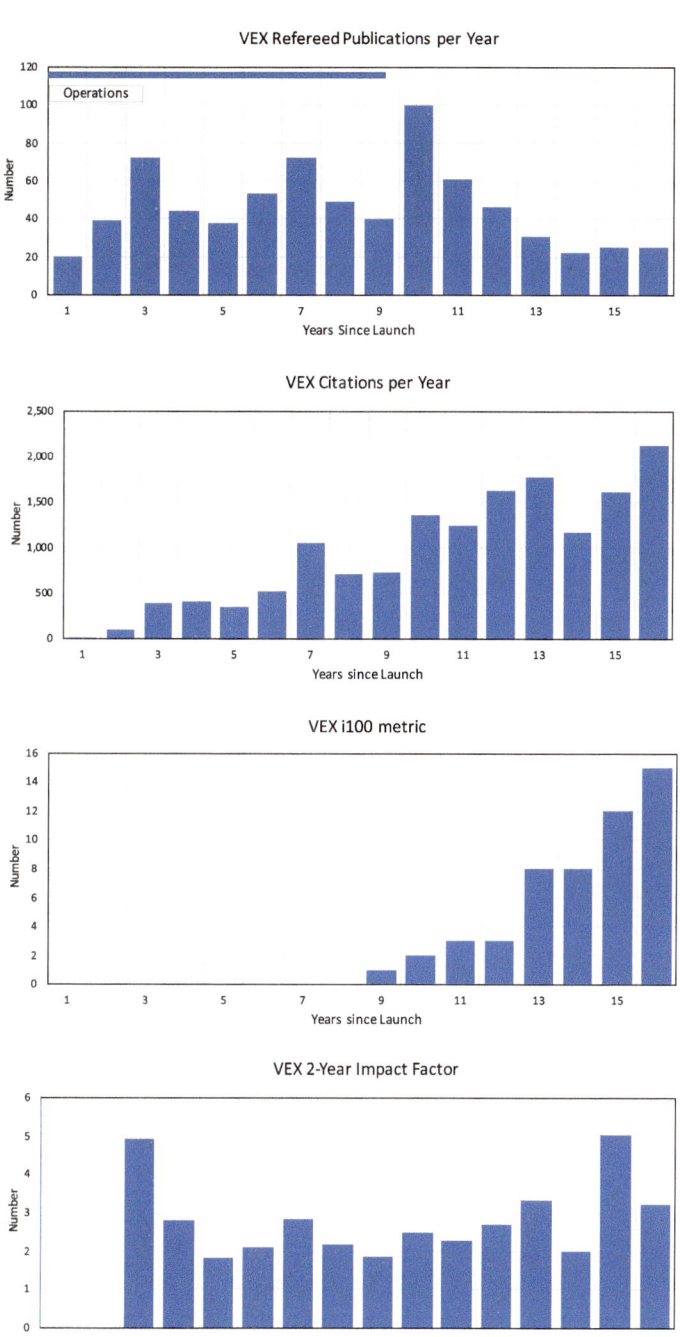

Fig. 2.23 VEX publication metrics

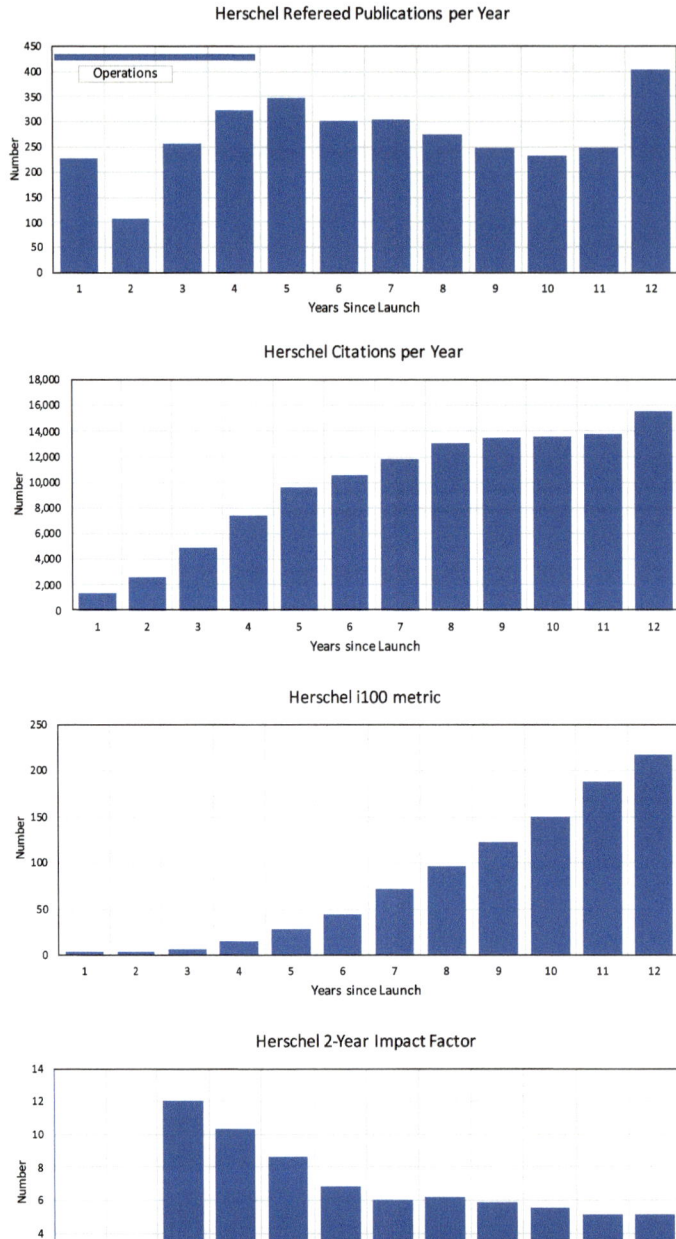

Fig. 2.24 Herschel publication metrics

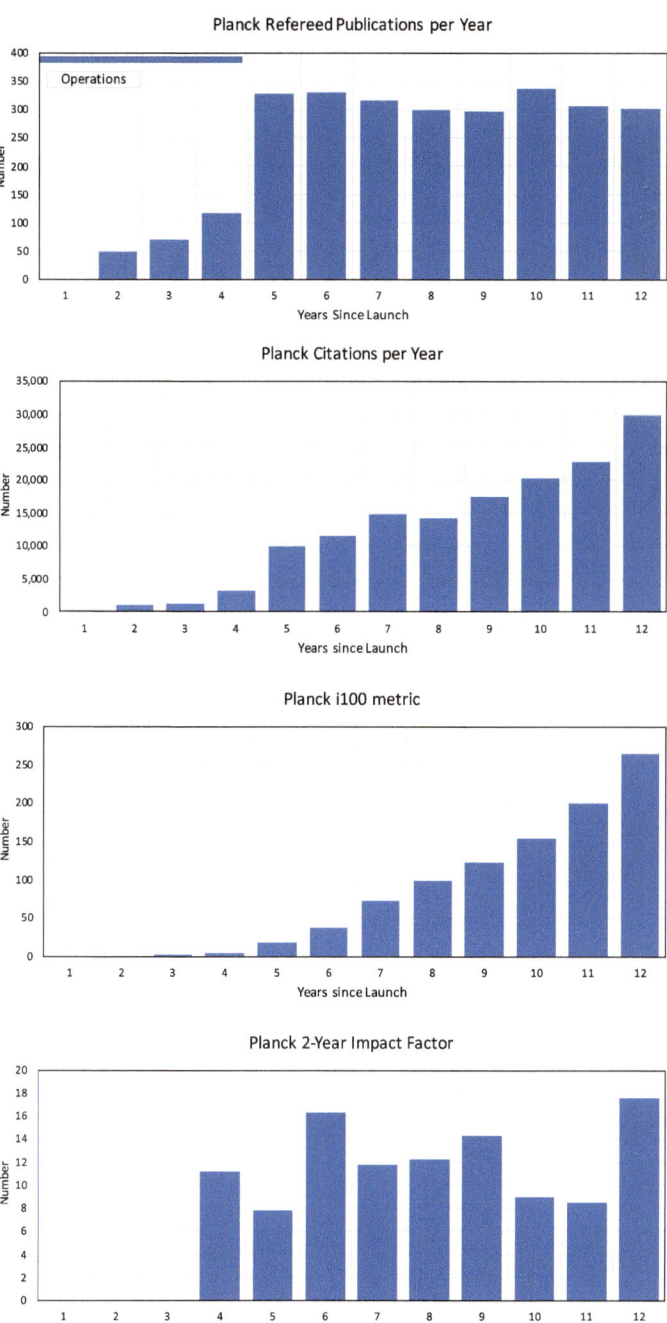

Fig. 2.25 Planck publication metrics

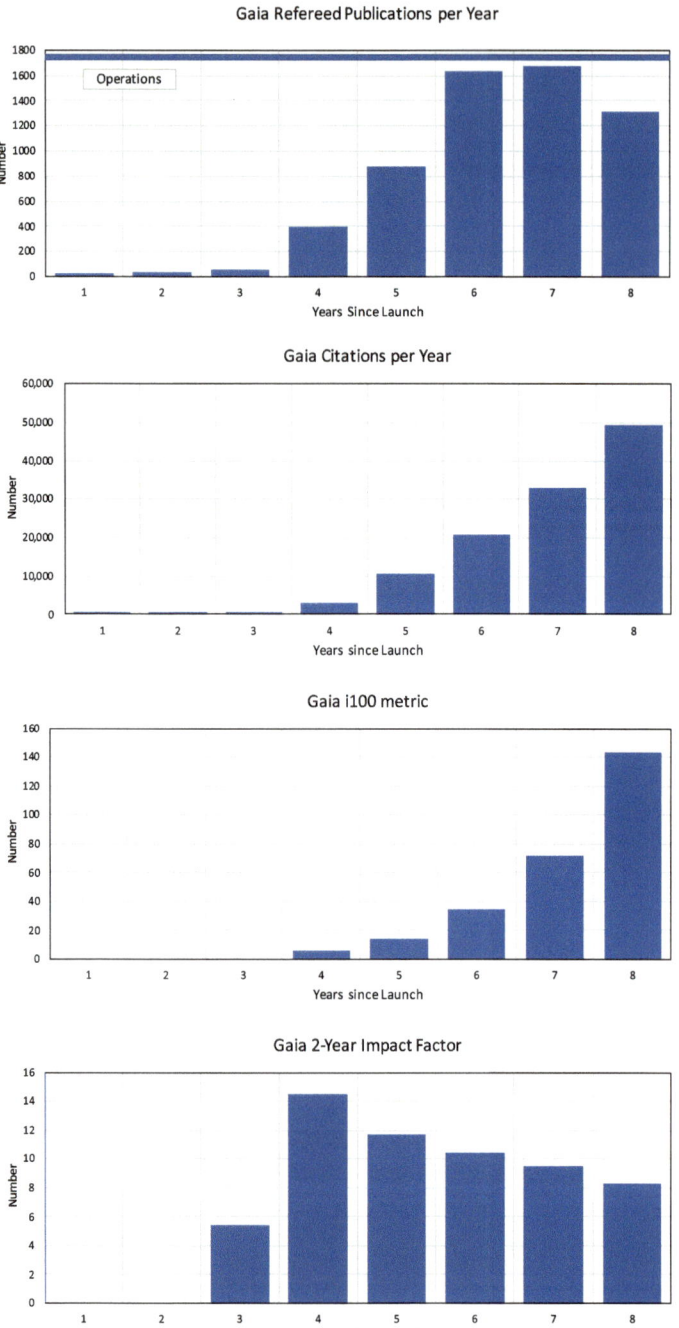

Fig. 2.26 Gaia publication metrics

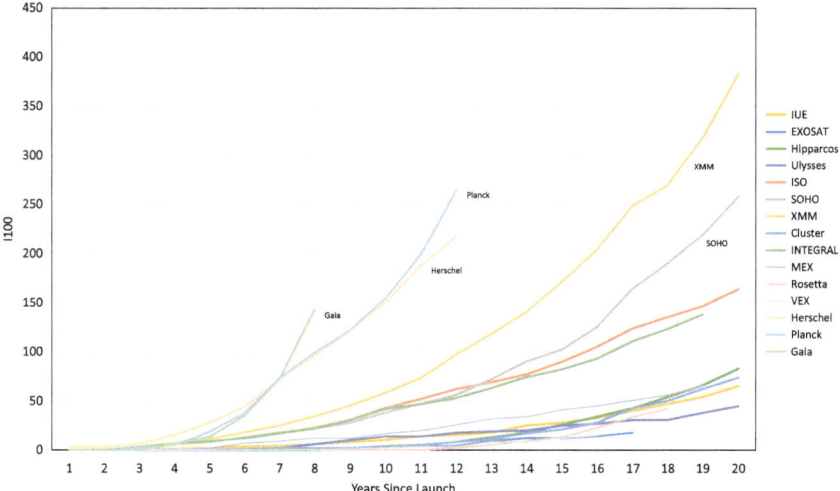

Fig. 2.27 Evolution of the i100 metric with years after launch for selected ESA-led missions. Different missions are represented by different coloured lines. Selected missions are individually labelled

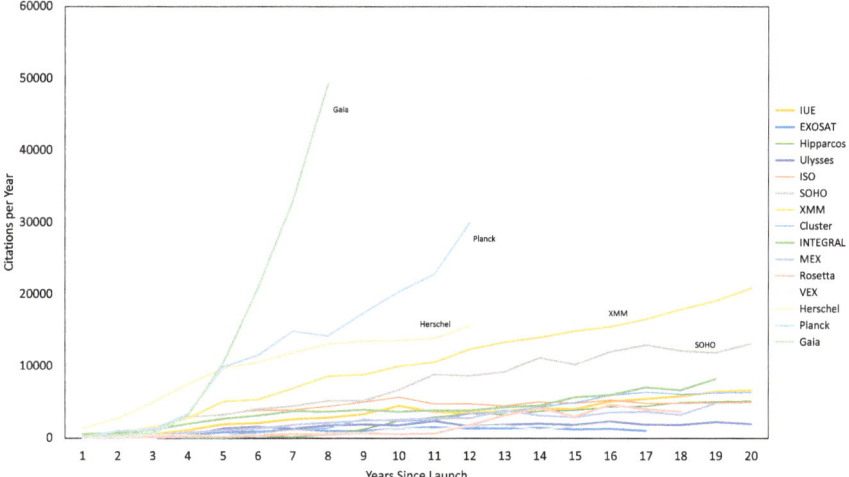

Fig. 2.28 Evolution of the number of citations with years after launch for selected ESA-led missions. Different missions are represented by different coloured lines. Selected missions are individually labelled

present in this section the statistics concerning such "archival publications" and their impact.

It is useful to point out that the phrases "archival publication", "archival paper", and "archival research" might be perceived differently in different communities. In the astronomy and heliophysics areas, science archives have been for decades the

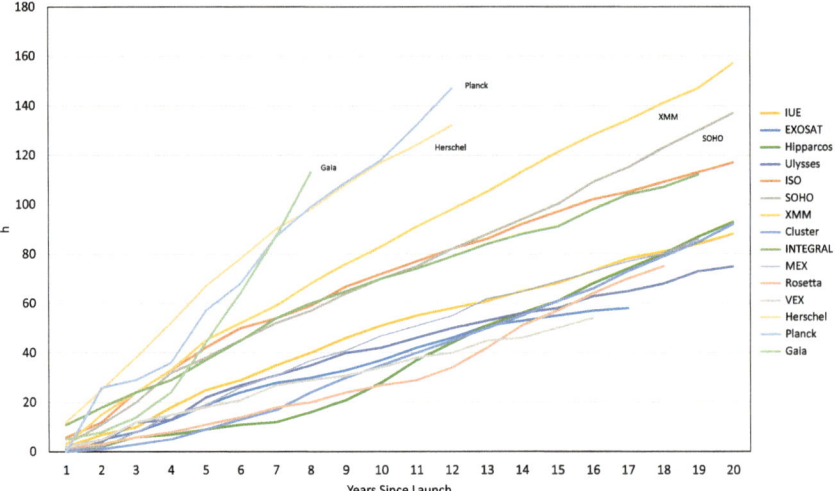

Fig. 2.29 Evolution of the h-index metric with years after launch for selected ESA-led missions. Different missions are represented by different coloured lines. Selected missions are individually labelled

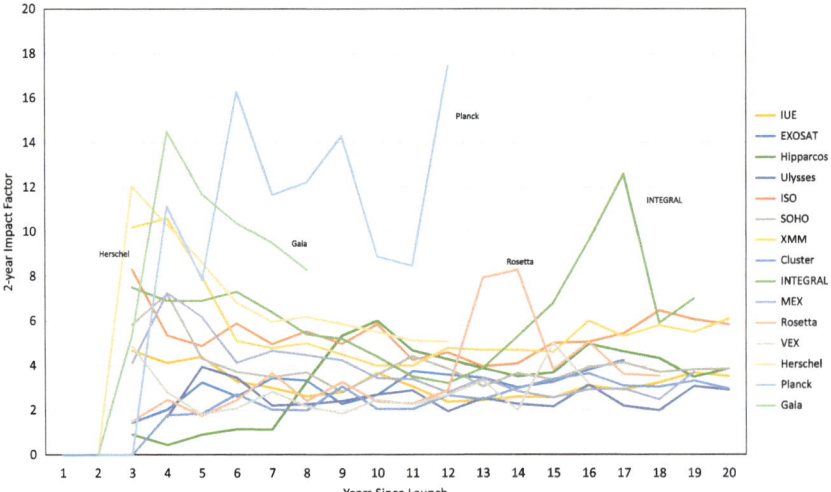

Fig. 2.30 Evolution of the 2-year impact factor with years after launch for selected ESA-led missions. Different missions are represented by different coloured lines. Selected missions are individually labelled

main tool to access scientific data, both for recent data, still covered by time-limited restricted access, and for public data, no longer subject to such restrictions. On the other hand, at ESA archives for Planetary Science were introduced about 20 years ago, and until recently data would still be delivered first directly to the teams of

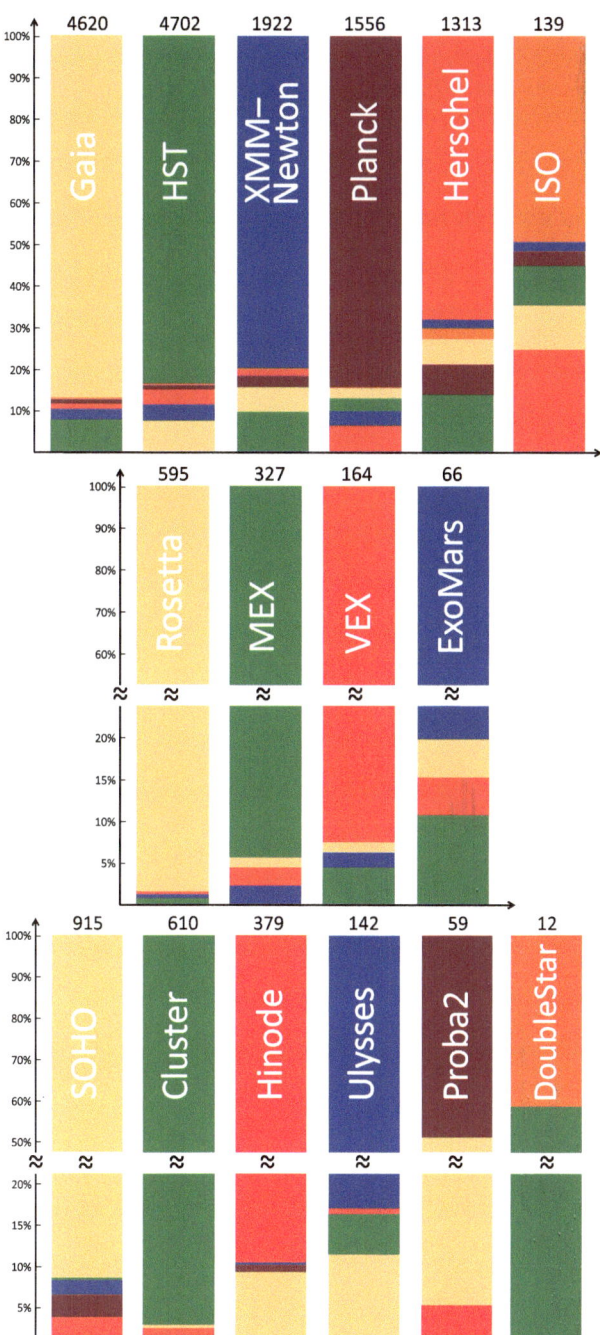

Fig. 2.31 Fraction of cross-mission papers for a number of representative missions over the period between 2016 and 2021. Astronomy missions are in the top panel, planetary missions in the middle, and heliophysics at the bottom. In each panel, a different colour is assigned to each mission and the total number of papers in that time frame is indicated at the top of each mission column. The typical fraction of cross-mission papers, weighted by the number of papers, is about 18% for astronomy, 5% for planetary science, and 9% for heliophysics

the Principal Investigator (PI)s and only later would they reach the ESA Planetary Science Archive. Here, the adjective "archival" will refer not so much to the place where the data are located (even though most of them are now physically in the ESA Space Science Archives) but rather to data that have already been used to address the original scientific questions posed by the researchers who first obtained or requested them.

In this light, in order to allow for meaningful comparisons across areas of research, we adopted a working definition of archival publications that is as uniform as possible for different missions. The adopted classification of a publication is based on the relationship between its authors and the scientists who led the research project generating the data used in that paper. The inspiring principle is that, in general, when a publication is written by scientists who built the instruments that produced those data or who were awarded observing time to collect those data, the publication is non-archival. If no such relationship exists, the publication is archival.

We searched for a definition applicable to all missions served by the ESA Space Science Archives, which can be roughly sorted in two main groups:

1. Observatory-type missions, where time-limited restricted access to the data of the on-board instruments is guaranteed to scientists who successfully qualify following an Announcement of Opportunity (AO).
2. Survey-type missions, where selected scientists provide on-board instruments to which they receive privileged, but time-limited, access during the mission.

Astronomy missions can fall into both groups, while Solar System missions typically belong to the second group. For some missions both possibilities exist. In all cases, however, after a period of restricted access, all data become available to all scientists in the community through the ESA Space Science Archives.

The definition of archival publication is as follows: a publication based on ESA mission data is considered fully archival when the list of authors does not include the PI of the proposal, instrument, or experiment producing those data. This definition draws on the classification so far adopted by ESA for XMM-Newton and Herschel publications, as well as by NASA for Chandra publications. The definition has been further adapted to cover all astronomy, heliophysics, and planetary science missions. Similarly, non-archival publication would be those that contain in the list of authors the name of the data or instrument PI. A third possibility is for partly-archival publications, which include data from more than one proposal, instrument or experiment without including all the corresponding PIs in the list of authors.

We point out that these definitions might slightly overestimate the number of archival publications, because they would not detect publications authored by some Co-Investigator (Co-I)s without including the PI. It would appear unlikely that the PI be specifically excluded, yet one does see this happening, particularly after several years of operations, when collaborations between team members become somewhat looser. Another, unrelated reason why the number of archival publications determined in this way might be slightly overestimated, particularly in the early years of operations, is the presence of publications not directly making use of the data (see Sect. 2.2). Examples are articles addressing technical aspects of the mission or of its instruments, scientific expectations, and simulations. Some of these publications do not include the PI as an author. These publications are conceptually easy to identify, but this cannot be done in an automated way and it requires the work of a scientifically trained librarian, which is not always possible for all ESA missions.[1] On the other hand, and perhaps more importantly, as long as the same definitions are applied consistently across all missions and over time, they will allow for meaningful comparisons and for the identification and monitoring of trends. We consider this an important element of this preliminary investigation.

For refereed publications based on XMM-Newton and Herschel data, the classification (archival versus non-archival) is simplified by the work done by members of the missions teams, who identified the specific data used in each paper and the corresponding observing proposal number. It is then simple to look for a match (or lack thereof) between the paper's authors and the proposal PI.[2]

The relevant statistics concerning these two missions are shown in Tables 2.14 and 2.15 and in graphical form in Figs. 2.32 and 2.33. The fraction of fully archival publications is different between the two missions, being higher for XMM-Newton most probably because of its longer history and, most importantly, its wider user community. In both cases, however, the fully archival publications show a steady growth over time, reaching ~35% for Herschel 10 years after launch and ~60% for XMM-Newton after 20 years of operations. For both missions, archival (fully and partly) publications are the majority: in recent years they exceed 60% of the total for Herschel and 80% for XMM-Newton.

The situation of publications based on HST data is also relevant in this context. In that case, however, the definition adopted by NASA also takes into account the role of Co-Is, so the statistics are not directly comparable with those of Herschel or XMM-Newton. Nevertheless, in the past ten years archival publications (fully and partly) have consistently represented over 60% of the total HST publications, showing an increase similar to that of the large ESA observatories.

For all other missions, the classification of publications as archival or non-archival is performed by checking the list of authors against the names of the PIs of onboard instruments and experiments. Applying the adopted definitions to the papers based on Planck data published in 2020 and earlier, results in 178 non-archival publications (i.e. those authored by the instrument PIs) out of 2652

---

[1] An alternative approach is the use of natural language processing and machine learning algorithms to identify and classify the publications automatically. This work is currently planned, as we briefly discuss in Sect. 2.7

[2] For the first few years of operations of XMM-Newton, a sizeable fraction of the papers cannot be classified with certainty because we lack specific knowledge of the members of the instrument teams. Furthermore, Guaranteed Time Observations and Target of Opportunity Observations were coarsely grouped in a small number of observing proposals, all under the name of the Project Scientist, with no information about the names of the scientists who were actually entrusted those data. Unfortunately, no digital records exist. Fortunately, these problems appear to be limited to the first few years of operations, and after 2006 the fraction of unclassified papers drops to close to zero.

Table 2.14 Temporal evolution of non-archival, partly-archival, and fully-archival publications based on XMM-Newton data. Values are given both as actual number of papers and in percentage

Year	Total	Non-archival		Partly-archival		Fully-archival		Unclassified	
2001	82	35	43%	5	6%	0	0%	42	51%
2002	120	75	62%	7	6%	6	5%	32	27%
2003	253	143	57%	50	20%	35	14%	25	10%
2004	351	147	42%	85	24%	93	26%	26	7%
2005	331	104	31%	86	26%	124	37%	17	5%
2006	369	93	25%	83	22%	186	50%	7	2%
2007	371	108	29%	78	21%	185	50%	0	0%
2008	339	85	25%	49	14%	205	60%	0	0%
2009	390	113	29%	62	16%	215	55%	0	0%
2010	375	85	23%	62	17%	228	61%	0	0%
2011	389	99	25%	83	21%	207	53%	0	0%
2012	339	82	24%	80	24%	177	52%	0	0%
2013	363	93	26%	73	20%	197	54%	0	0%
2014	359	97	27%	70	19%	192	53%	0	0%
2015	325	84	26%	49	15%	192	59%	0	0%
2016	367	94	26%	91	25%	182	50%	0	0%
2017	387	88	23%	80	21%	219	57%	0	0%
2018	462	105	23%	85	18%	272	59%	0	0%
2019	443	91	21%	87	20%	265	60%	0	0%
2020	387	75	19%	77	20%	235	61%	0	0%
Total	6802	1896	28%	1342	20%	3415	50%	149	2%

Table 2.15 Temporal evolution of non-archival, partly-archival, and fully-archival publications based on Herschel data. Values are given both as actual number of publications and in percentage

Year	Total	Non-archival		Partly-archival		Fully-archival		Unclassified	
2010	228	4	2%	200	88%	20	9%	4	2%
2011	109	1	1%	100	92%	6	6%	2	2%
2012	256	10	4%	220	86%	19	7%	7	3%
2013	323	6	2%	269	83%	43	13%	5	2%
2014	347	8	2%	278	80%	54	16%	7	2%
2015	301	5	2%	230	76%	62	21%	4	1%
2016	305	3	1%	215	70%	81	27%	6	2%
2017	276	2	1%	196	71%	71	26%	7	3%
2018	251	12	5%	150	60%	84	33%	5	2%
2019	246	86	35%	75	30%	78	32%	7	3%
Total	2642	137	5%	1933	73%	518	20%	54	2%

publications in total. The evolution over time of archival and non-archival Planck publications is shown in Fig. 2.34. This implies a fraction of archival publications exceeding 93%. Alternatively, one could consider as non-archival publications those authored by the Planck Collaboration, which include the phrase "Planck Collaboration" in the list of authors. These amount to 159 articles in the same period and result in a slightly higher fraction of archival publications, namely almost 94%.

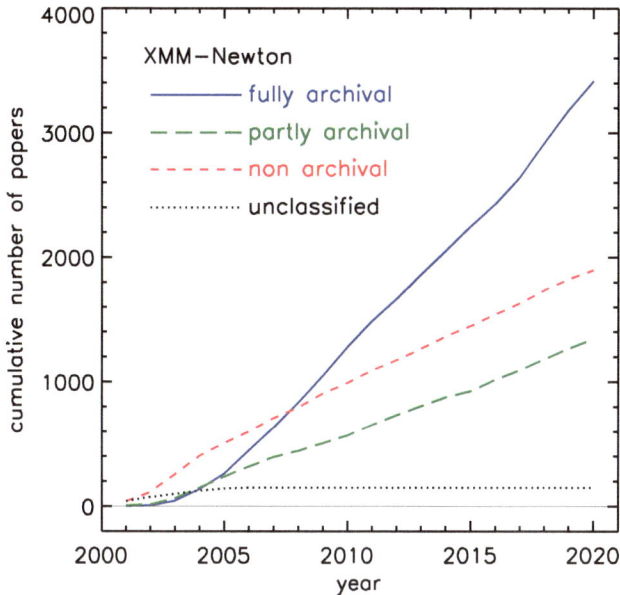

Fig. 2.32 Cumulative distribution of XMM-Newton refereed papers as a function of time for different categories (see legend)

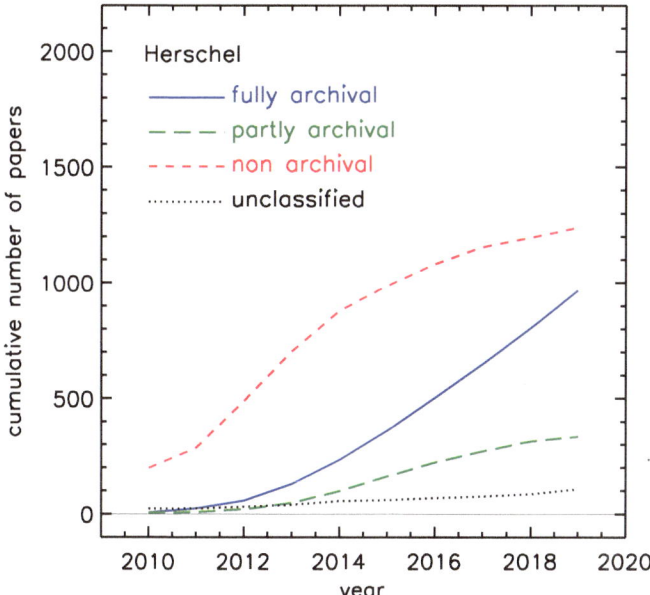

Fig. 2.33 Cumulative distribution of Herschel refereed papers as a function of time for different categories (see legend)

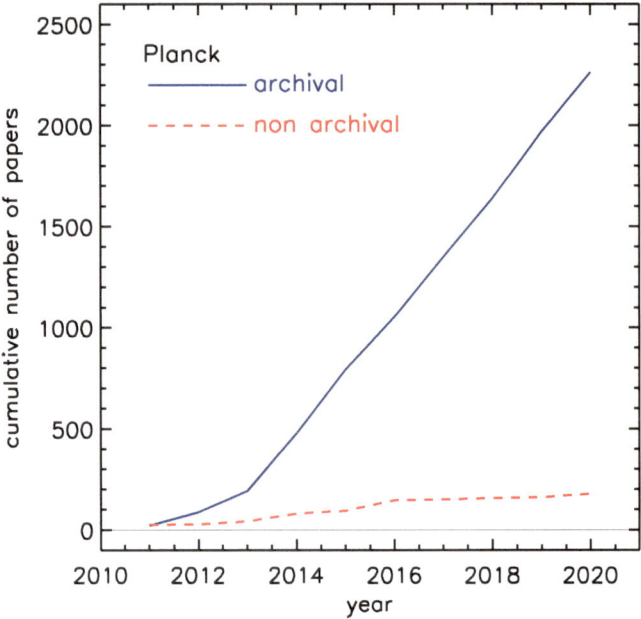

Fig. 2.34 Cumulative distribution of Planck refereed publications as a function of time. Archival and non-archival publications are shown (see legend)

Regardless of the exact metrics adopted, there is no doubt about the overwhelmingly archival nature of the literature based on Planck data.

The situation of Gaia, one of the most prolific space science missions so far, is even more extreme: in this case there is no time-limited restricted access to the data and all Gaia Data Releases are immediately available to the entire community. Therefore, all publications based on Gaia data are by definition fully archival. The publication numbers in Table 2.1 above include a small amount of calibration and processing papers too, but in the case of Gaia those represent less than 2% of the total.

Archival publications represent a sizeable fraction of the literature also for ESA Solar System missions (planetary science and heliophysics). Also here is the classification based on the existence of a match (or lack thereof) between the list of authors in a publication and the names of the PIs of the instrument or experiments onboard those missions. As representative examples, we show in Figs. 2.35 and 2.36 the cases of MEX and Rosetta, respectively. While for Rosetta the proportion of archival and non-archival publications has been relatively similar throughout the years, for MEX the number of archival publications has consistently surpassed that of non-archival publications starting in 2016. The preponderance of archival publications is even more marked for heliophysics missions, as we show in Fig. 2.37 for Cluster and Fig. 2.38 for SOHO, undoubtedly also due to the very long operational phase of the latter mission. With a total fraction of over 85 % of archival

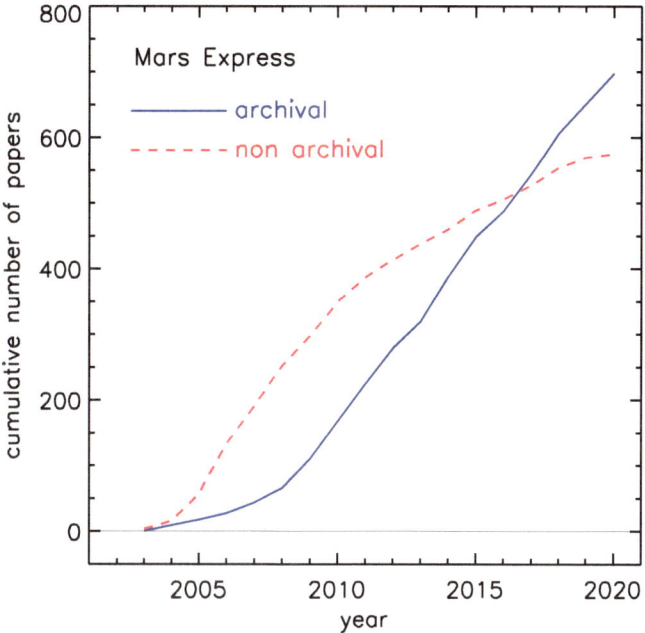

Fig. 2.35 Cumulative distribution of MEX refereed publications as a function of time. Archival and non-archival publications are shown (see legend)

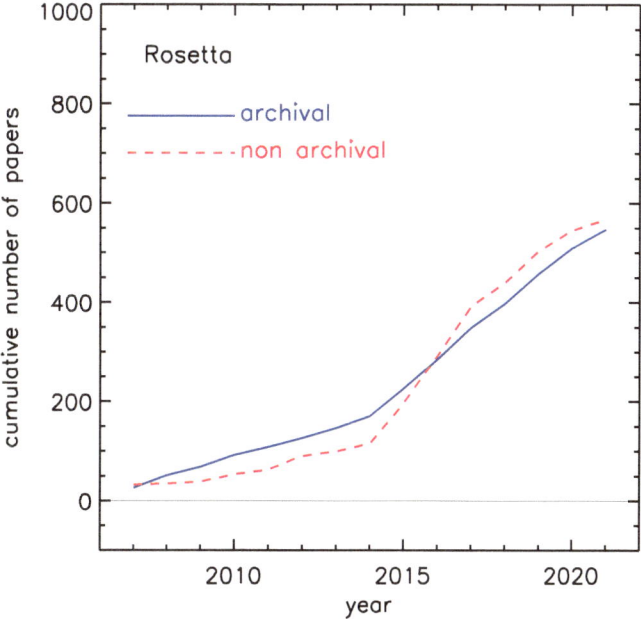

Fig. 2.36 Cumulative distribution of Rosetta refereed publications as a function of time. Archival and non-archival publications are shown (see legend)

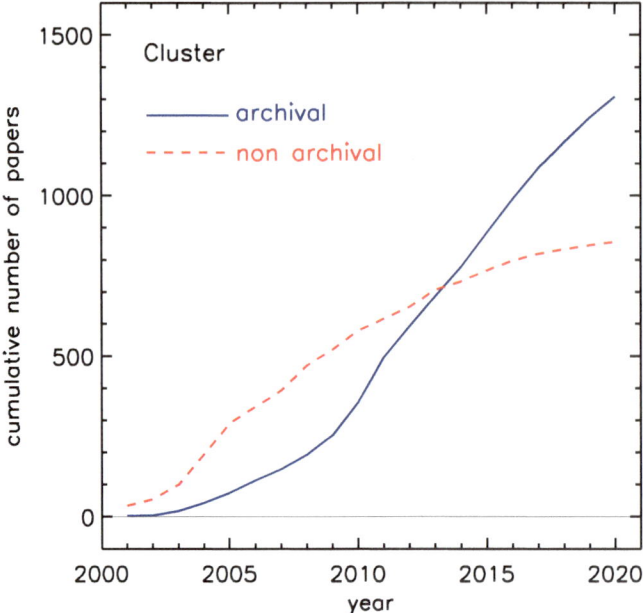

Fig. 2.37 Cumulative distribution of Cluster refereed publications as a function of time. Archival and non-archival publications are shown (see legend)

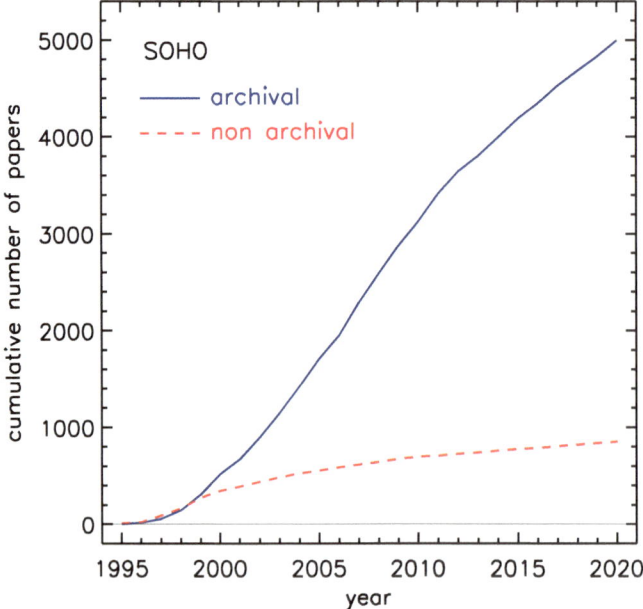

Fig. 2.38 Cumulative distribution of SOHO refereed publications as a function of time. Archival and non-archival publications are shown (see legend)

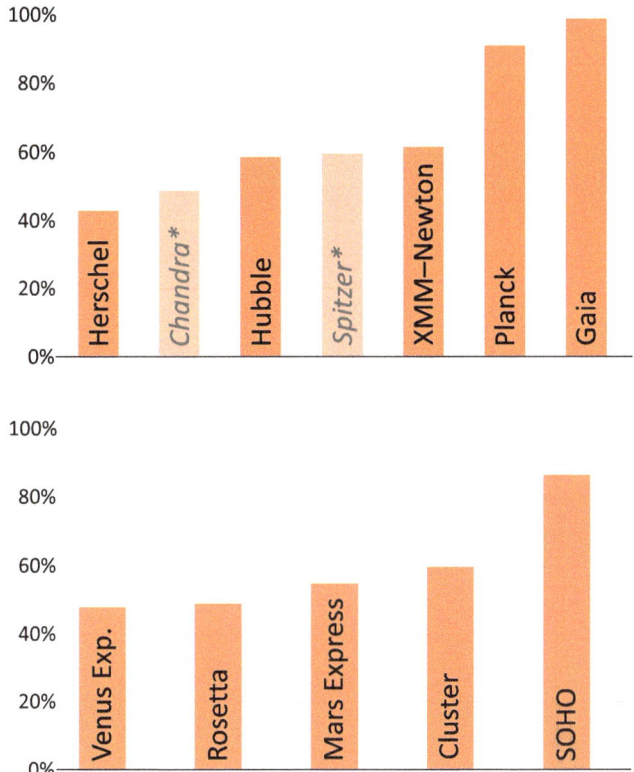

Fig. 2.39 Cumulative fractions of archival papers for a sample of ESA Astronomy (top) and Solar System (bottom) missions. (*) Data about Chandra and Spitzer are from [25]

papers by 2020, SOHO is after Gaia and Planck the mission with the highest fraction of archival papers and, together with Planck, it is one of the missions with the fastest growth of archival papers over time.

The cumulative fractions of archival papers for a sample of ESA Astronomy and Solar System missions are shown in Fig. 2.39. The statistics cover different time spans and the various missions have rather different lifetimes, so it is difficult to directly compare the mission to one another. However, the purpose of this figure is to show that ESA missions in general have produced a sizeable fraction of archival papers, which appears to be fully in line with that generated by NASA missions. For reference, we have included in the graph the statistics about two of NASA's Great Observatories, namely Chandra and Spitzer (from [25]).

A relevant question to ask is whether the impact of archival papers is similar to that of non-archival publications. To address this question, we explored the temporal evolution of the number of citations of both types of papers from four representative missions: an astronomical observatory (XMM-Newton), an astronomical survey mission (Planck), a planetary mission (MEX), and a heliophysics mission (Cluster).

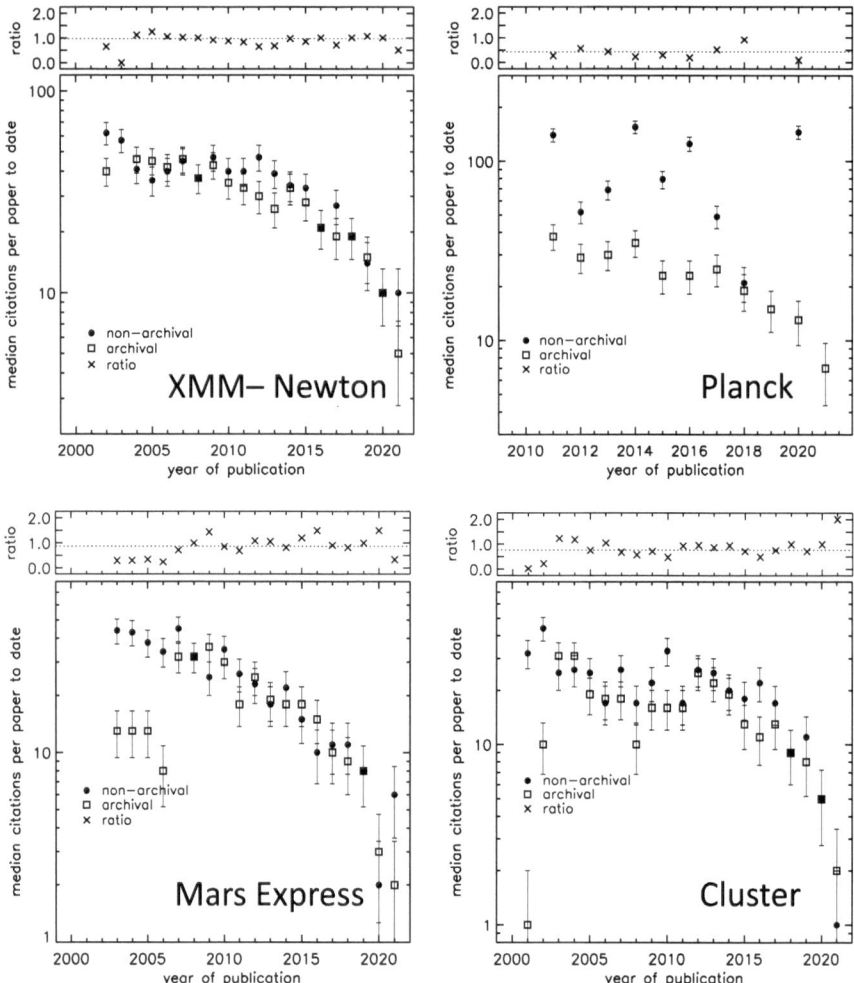

Fig. 2.40 Median number of citations to date for a sample of four representative ESA Science Programme missions. Non-archival papers are indicated with dots, archival papers with squares. Error bars assume Poisson statistics. Crosses in the upper panels display the ratio of archival and non-archival papers

We show in the lower panels of Fig. 2.40 the median number of citations that papers written over the years have received until the date the calculations were made (late 2022). Dots and squares are used, respectively, for non-archival and archival papers and error bars assume Poisson statistics.

Understandably, less recent papers have had more time to accumulate citations over the years, so the general decreasing trend is expected (although this is less obvious in the case of Planck's publications, as discussed below). Not surprisingly,

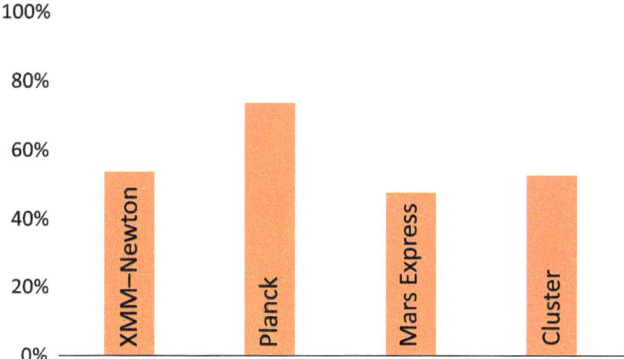

Fig. 2.41 Cumulative fractions of citations to archival papers for the same missions as in Fig. 2.40

in the first years of operations non-archival papers receive typically more citations because of the novel data that they present and because it takes some time before those data become available to other scientists not originally involved in the experiments or proposals. However, the difference becomes progressively smaller and in some instances archival papers receive on average more citations than non-archival papers. The top panels of Fig. 2.40 show the ratio of archival and non-archival papers as a function of time, where the dotted lines indicate the median value of the ratio, which is 0.97 for XMM-Newton, 0.43 for Planck, 0.86 for MEX, and 0.76 for Cluster. This comparison shows that, in general, archival papers have an impact only slightly smaller than that of non-archival works, and the difference decreases further after the first few years of operations.

The large difference between Planck and the other three missions can probably be understood considering that the Planck non-archival papers are the 159 refereed articles authored by the Planck Collaboration. Many of those provide catalogues and fundamental cosmological parameters that have become primary references in the field. The fact that they are highly cited attests to the major contribution that the work done by the Planck PIs and Planck Collaboration has given to cosmology. But the impact, success, and contribution of this and other missions is also measured by the total number of citations, which in the case of Planck is clearly dominated by non-archival papers. As we show in Fig.2.41, about 3/4 of the citations that Planck publications receive are to archival papers, while for the other missions in the sample archival and non-archival papers receive a similar number of citations.

In summary, this analysis has revealed that non-archival and archival papers contribute in similar ways to the visibility of the missions and, more importantly, to the advancement of science. For this to be possible, it is crucial that all the data from the missions, including the calibration and associated documentation, are shared with the community worldwide and preserved, including when the missions are no longer in operation.

2.7   Conclusions

We have examined almost half a century of scientific publications resulting from data generated by the ESA Science Programme's missions. The number of papers has increased significantly over time, matching the growth and reach of the ESA-led missions, from less than ten papers per year in the mid-to-late 1970s, when COS-B was the only mission in the programme, to over 4000 publications per year at the end of 2021, when the ESA Science Programme counted 11 active missions and 14 more in their legacy phases, all of which are still contributing to new discoveries.

At first sight, one might attribute the increase in the number of publications from ESA's Science Programme missions to the general increase in scientific publications across all areas of space science over the past decades. Indeed, the collective number of papers in the fields of astronomy, planetary science, and heliophysics, as tracked by the ADS, has grown from ∼20,000 publications per year in 2000, to ∼26,000 in 2010, and ∼35,000 in 2020. It is interesting that the fraction of those papers that make use of data from ESA mission has also grown with time, but at an even faster rate: In 2000, ESA mission papers accounted for 4.1% of all the astronomy, planetary science and heliophysics papers published that year; in 2014 the fraction had almost doubled, to 7.6%, and in 2021 that fraction is about three times higher, namely 11.5% for the ESA-led Science Programme missions alone, and 15.2% when the partner-led missions are also included. Thus, in a world in which the number of scientific papers keeps increasing, the fraction of papers based on ESA's Science Programme missions increases even more rapidly, witnessing an ever growing contribution by these missions to the advancement of science.

Besides the steady growth in the number of publications, our analysis shows that the scientific productivity of the missions remains high even after the end of operations. In fact, we have discovered that, in all fields of space science, archival papers represent the majority of the publications based on ESA data, and that their impact, as judged by the number of citations that they receive, is typically as large as that of non-archival papers. This highlights the importance of maintaining the missions' data available for the scientific community to explore and investigate well beyond the end of mission operations, potentially for as long as the data remain relevant. This is the role that ESA's Science Directorate has entrusted to the ESA Space Science Archives.

Our analysis shows that the geographical distribution of the publications appears generally in line with the size of the communities in the various countries. For the ESA Member States, it roughly scales with the financial contribution of each country to the ESA Science Programme, which in turn is linked to the GDP of each State. Some countries, both large and small, stand out in that their scientific communities appear to be very effective at turning mission data into scientific discoveries. Furthermore, the cases of some of the smaller countries (e.g., Hungary, Finland, Czech Republic) show that scientific research and investigations, all of which lead to scientific papers, are an effective way in which countries can become involved in space science, even before they have developed significant space technology infrastructure.

We close with some considerations about the future of this research. This investigation was possible thanks to the meticulous work done over the years by the many ESA Project and Contact Scientists who regularly survey, parse, and read the relevant scientific literature to identify papers that make use of data from the ESA Science Programme missions. No matter how refreshing and interesting it might be for scientists to learn about new discoveries and advancements in their fields, these tasks are labour intensive, particularly for very productive missions. Utilising advancements in the field of machine learning, the ESA Science Directorate is transitioning from manual identification and selection of literature papers to a semi-automated approach, whereby machine learning and natural language processing algorithms are used to sift through the literature and identify papers likely to make use of data from the ESA Science Programme missions and that fulfil the conditions indicated above. This is simplified by the availability of extensive training sets made up of the thousands of literature papers that were inspected by a scientist and eventually accepted or discarded. The papers automatically identified with these tools are then still validated by the Project and Contact Scientists. Preliminary analysis has shown that this might introduce small uncertainties in the number of papers, at the level of 10%, which are however comparable with those encountered when different scientists search and identify papers for the same mission. This holds the promise of even more detailed investigations, not just on the sheer number and impact of the papers, but also for instance on the types of data and on the methods that are used in the papers. We plan to report on the continuation of this study in a future work.

Acknowledgments We thank the ESA Project and Contact Scientists for their careful curation of the publication libraries. We also thank Edwin Henneken from the NASA Astrophysics Data System Bibliographic Services for his invaluable support. This research was supported by the International Space Science Institute (ISSI) in Bern, through the ISSI Working Group project "The Scientific Performance of ESA's Science Missions".

References

1. L. Scarsi, K. Bennett, G.F. Bignami, et al., The Cos-B experiment and mission, in *Recent Advances in Gamma-Ray Astronomy*, ed. by R.D. Wills, B. Battrick, vol. 124 (ESA Special Publication, 1977), p. 3
2. A. Accomazzi, E.A. Henneken, C. Erdmann, et al., Telescope bibliographies: an essential component of archival data management and operations, in *Observatory Operations: Strategies, Processes, and Systems IV*, ed. by A.B. Peck, R.L. Seaman, & F. Comeron. Society of Photo–Optical Instrumentation Engineers (SPIE) Conference Series, vol. 8448 (2012), p. 84480

3. A. Accomazzi, M.J. Kurtz, E.A. Henneken, et al., ADS: the next generation search platform, in *Open Science at the Frontiers of Librarianship*, ed. by A. Holl, S. Lesteven, D. Dietrich, et al. Astronomical Society of the Pacific Conference Series, vol. 492 (2015), p. 189

4. E.A. Henneken, A. Accomazzi, M.J. Kurtz, et al., Computing and using metrics in the ADS, in *Open Science at the Frontiers of Librarianship*, ed. by A. Holl, S. Lesteven, D. Dietrich, et al. Astronomical Society of the Pacific Conference Series, vol. 492 (2015), p. 80

5. A.H. Rots, S.L. Winkelman, G.E. Becker, Chandra publication statistics. Publ. Astr. Soc. Paci. **124**(914), 391 (2012)

6. F. Jansen, D. Lumb, B. Altieri, et al., XMM-Newton observatory. I. The spacecraft and operations. Astron. Astrophys. **365**, L1–L6 (2001)

7. B. Fleck, V. Domingo, A.I. Poland, The SOHO mission. Solar Phys. **162**(1), 1–531 (1995)

8. M.A.C. Perryman, The GAIA mission, in *GAIA Spectroscopy: Science and Technology*, ed. by U. Munari. Astronomical Society of the Pacific Conference Series, vol. 298 (Society of the Pacific Astronomical, San Fransisco, 2003), p. 3

9. T. Owen, The Cassini mission, in *NASA Conference Publication*, vol. 2441 (1986), pp. 231–237

10. M.A.C. Perryman, Hipparcos: astrometry from space. Nature **340**(6229), 111–116 (1989)

11. C.P. Escoubet, M. Fehringer, M. Goldstein, Introduction: the cluster mission. Ann. Geophys. **19**, 1197–1200 (2001)

12. J.E. Hirsch, An index to quantify an individual's scientific research output. Proc. Natl. Acad. Sci. **102**(46), 16569–16572 (2005)

13. P. Benvenuti, The IUE mission. Memoir. Ital. Astron. Soc. **54**, 359 (1983)

14. G.L. Pilbratt, J.R. Riedinger, T. Passvogel, et al., Herschel space observatory. An ESA facility for far-infrared and submillimetre astronomy. Astron. Astrophys. **518**, L1 (2010)

15. J.A. Tauber, N. Mandolesi, J.L. Puget, et al., Planck pre-launch status: the planck mission. Astron. Astrophys. **520**, A1 (2010)

16. K.P. Wenzel, R.G. Marsden, D.E. Page, et al., The ULYSSES mission. Astron. Astrophys. Suppl. **92**, 207 (1992)

17. B.G. Taylor, R.D. Andresen, A. Peacock, et al., The EXOSAT mission. Space Sci. Rev. **30**(1–4), 479–494 (1981)

18. E. Kuulkers, C. Ferrigno, P. Kretschmar, et al., INTEGRAL reloaded: spacecraft, instruments and ground system. New Astron. Rev. **93**, 101629 (2021)

19. J.L. Vago, B. Gardini, P. Baglioni, et al., Science objectives of ESA's ExoMars mission, in *European Planetary Science Congress 2006* (2006), p. 76

20. J. Schneider, M. Auvergne, A. Baglin, et al., The COROT mission: from structure of stars to origin of planetary systems, in *Origins*, ed. by C.E. Woodward, J.M. Shull, A. Thronson Harley, Jr. Astronomical Society of the Pacific Conference Series, vol. 148 (1998), p. 298

21. J. Bergé, P. Touboul, M. Rodrigues, et al., *Status of MICROSCOPE, a Mission to Test the Equivalence Principle in Space*. Journal of Physics Conference Series, vol. 610 (2015), p. 012009

22. S. Tsuneta, The hinode mission, in *First Results From Hinode*, ed. by S.A. Matthews, J.M. Davis, L.K. Harra. Astronomical Society of the Pacific Conference Series, vol. 397 (2008), p. 3

23. K.H. Glassmeier, H. Boehnhardt, D. Koschny, et al., The Rosetta Mission: Flying Towards the Origin of the Solar System. Space Sci. Rev. **128**(1-4), 1–21 (2007)

24. M.F. Kessler, J.A. Steinz, M.E. Anderegg, et al., The infrared space observatory (ISO) mission. Astron. Astrophys. **315**(2), L27 (1996)

25. J. Peek, V. Desai, R.L. White, et al., Robust archives maximize scientific accessibility (2019). https://arxiv.org/abs/1907.06234

Chapter 3
Payload Provision to the ESA Science Programme

John Zarnecki and Arvind Parmar

Abstract We have collected data pertaining to the Principal Investigators (PIs), and co-PIs (where appropriate) for all ESA-led Science Directorate missions since the first such launch, namely of COS-B in 1975. For a total of 28 missions (including 4 in preparation awaiting launch), 437 individuals have been recorded along with their institution, location, "academic age" and gender. We have correlated the number of PIs by country with the financial contribution of those countries to the ESA Science programme. We have also investigated issues associated with age and gender of the PIs. As a result of these analyses, we make suggestions for actions which ESA and its Member States may wish to consider with the aim of encouraging equity and diversity while still placing scientific excellence as the overarching goal.

Keywords ESA · European Space Agency · ESA Science Programme · Member State · Member state payload provision · Payload complexity · Principal Investigator · Co-Principal Investigator · Gender · Gender balance · Proposer institute country · Proposer academic age

3.1 Introduction

European Space Agency (ESA)'s Science Directorate missions have, over the last five decades, followed a variety of formats. However, two distinct patterns can be discerned from the large number of missions flown. At one end of the spectrum there are missions which incorporate a collection of separate instruments, or payloads, generally functioning independently of each other, at least from a technical perspective, although very often the scientific aims and objectives are

J. Zarnecki (✉)
The Open University, Milton Keynes, UK
e-mail: J.C.Zarnecki@open.ac.uk

A. Parmar
Department of Space and Climate Physics, MSSL/UCL, Dorking, UK
e-mail: arvind.parmar@ucl.ac.uk

© The Author(s) 2025
A. Parmar et al., *ESA Science Programme Missions*, ISSI Scientific Report Series 18,
https://doi.org/10.1007/978-3-031-69004-4_3

interrelated, sometimes very closely. Very occasionally there may be a closer relationship between two such instruments—an example of such a dependency occurred on the Huygens Probe [1] where the Aerosol Collector and Pyrolyser (ACP) [2] instrument collected aerosol particles in the atmosphere of Titan and passed these to a separate instrument, the Gas Chromatograph Mass Spectrometer (GCMS) [3] for subsequent analysis. However, even in such cases, scientific instruments maintain their separate technical, scientific and organisational identities. Such collections of instruments predominantly, or even exclusively, occur with missions dedicated to the exploration of the Solar System and typically involve travelling to the target of interest and making a series of observations remotely, often in close proximity to the target, or in-situ. Examples of the former include missions such as Venus Express Mission (VEX) [4], BepiColombo [5] and Solar and Heliospheric Observatory (SOHO) [6] while the latter encompasses such missions as Rosetta [7] and Huygens [1]. Occasionally, a mission will include instruments for both in-situ and remote observations such as Giotto [8] and Solar Orbiter [9].

Alternatively, there are observatory missions. These typically involve a telescope-type payload with one or more instruments deployed to analyse the light collected in whatever region of the electromagnetic spectrum that is under investigation. A variant is to have a small collection of, often co-aligned, instruments to study the same target or region of sky. Unlike the category previously described, these missions are almost exclusively of an astrophysical nature, normally operating in those parts of the electromagnetic spectrum inaccessible from the Earth's surface and using space as a convenient location from which to collect and analyse the relevant information. Examples of such missions include INTEGRAL (Gamma-ray) [10], XMM-Newton (X-ray) [11] and Herschel (infra-red and sub-mm) [12].

Another type of mission are those such as Gaia [13] where the payload was procured directly by ESA and the scientific community is responsible for the scientific data processing—a task comparable in scope to delivering a major instrument in this case. It should be noted that instrument consortia often provide parts of the scientific ground segment of ESA's Science Directorate missions.

Whichever the mission type, instruments are usually provided by consortia of scientists from one or more countries both from the ESA Member States and beyond. Instruments vary in size and complexity from simple through to extremely large and complex costing many tens of million of euros. Instruments are usually selected competitively guided by the ESA Science Advisory Structure (see Chap. 1). This selection process can be complex, involving a whole range of factors including scientific and technical excellence, organisational and managerial feasibility and financial viability. Issues of industrial return, and international balance can also play a significant role. In the case where the mission involves collaboration with other agencies (i.e., National Aeronautics and Space Administration (NASA), Japanese Aerospace Exploration Agency (JAXA), China National Space Agency (CNSA), etc), then the national balance between the relevant agencies will also come into play. Selection of an instrument consortium implies a degree of recognition and trust in that consortium and in the consortium leader, or leaders, who will by implication be regarded as having significant scientific standing in the scientific area of the particular mission as well as the necessary scientific, technical, managerial and organisational skills required to bring the instrument to fruition. Thus, selection as a Principal Investigator (PI), as the consortium leader is usually referred to, undoubtedly conveys a strong degree of recognition within the space science community and the wider research domain.

Since the process of payload selection has been in progress for approaching 50 years in a relatively homogeneous way, this offers the opportunity for an analysis of the results of such selections from the perspective of at least the nation of the PI's host institution and other factors such as age and gender, although in the latter case, the relative paucity of numbers makes this more difficult. This is an area of investigation that has been relatively neglected to date yet has the potential to provide interesting and potentially instructive insights into the process, its outcomes and implications. [14] analysed the participation of women scientists in 10 ESA Solar System missions selected between 1980 and 2012 finding between 4 and 25% participation with several missions with no women PIs. Biases, both conscious and unconscious, may be at play in the process but that is not the underlying premise of this study. The whole process of payload selection is complex and depends on many interrelated factors. Understanding these and improving the process, if that is indeed possible, partly depends on data collection and analysis of the type that is presented here. We do not claim any definitive answers at this stage but believe that analysis of this type may help to understand better some of the factors at play and to improve the process where possible.

We note that not all the current Member States have been members since the start of ESA. The eleven founding Member States of Belgium, Denmark, France, West Germany, Ireland, Italy, the Netherlands, Spain, Sweden, Switzerland, and the United Kingdom joined in 1975. These were followed by Austria in 1984, Norway in 1985, Finland in 1994, Portugal in 2000, Greece and Luxembourg in 2004, the Czech Republic in 2008, Romania in 2011, Poland in 2012, and Estonia and Hungary in 2015. Science Programme payload selections are normally open worldwide, including countries that are not yet Member States. However, for the non-founding Member States, their first payload PI or co-PI was usually appointed some years after becoming members of ESA. Two exceptions are Finland which joined ESA in 1995 and had a payload PI on SOHO whose payload was selected in 1989 and Hungary which became a member in 2015 and had a PI on Rosetta whose payload was selected in 1993 (see Table 3.1).

3.2   Methodology

Data were collected on 28 ESA-led missions in the ESA Science Programme from a variety of sources including mission web sites, press releases, papers published in the scientific literature and the NASA Space Science Data Coordinated Archive (nssdc.gsfc.nasa.gov). The interval covered started with COS-B [15] selected in 1969 and launched in 1975 and continued to include those missions where the

Table 3.1 The number of originally selected PIs and co-PIs contributing to payload or science operations from each ESA Member State. % Fin. Contrib. is the percentage financial contribution between 2000 and 2021 of Member States to the Science Programme. Class refers to Small/Smart, Medium or Large sized mission class in the ESA Science Programme. Year is the year of mission or payload selection. Ratio is the fraction of publications from a Member State divided by the financial contribution to the Science Programme from that Member State. The Weighted Ratios include factors for average co-PI contributions (assumed to be one half that of a PI), mission class (factors of 2.0, 1.0 and 0.3 were applied for Large, Medium and Small/Smart missions), and inversely to the total number of PIs and co-PIs that a mission has (including from non-ESA Member States)

Mission	Class	Year	AT	BE	CH	CZ	DE	DK	EE	ES	FI	FR	GB	GR	HU	IE	IT	LU	NL	NO	PL	PT	RO	SE	Total ESA MS
COS-B	M	1969					1					1					2		1						5
GEOS	M	1970			1	4	1				3	1	1			1	1								12
Ulysses	M	1977			1		6					1	1			1	1								9
Giotto	M	1980			1		4					2	2												11
Hipparcos	M	1980						1				2							1					1	4
ISO	L	1983					1					1	1				1								4
Huygens	M	1988			1		1					1	2				1								5
SOHO	M	1989					4				1	3	2				1			1					13
Cluster	M	1989					2					3	3												10
XMM-Newton	L	1991	1				1						2				1		1					1	5
Rosetta	L	1993	2		1		11				1	6	3		2		3							2	31
INTEGRAL	M	1993			1		1	1		1		3					2								9

Planck	M	1996						1				2					2							5
Herschel	L	1998		1						2		2	1				1							7
SMART-1	S	1999			1		2			3			1						2					9
MEX	M	1999								2	2	2							2	2		8
Gaia	M	2000											1			1					2
BepiColombo	L	2000	2		1		4		3	6	4		1		7	1	3			3	31
Solar orbiter	M	2000		1	2	1	5			2	4	5			2			1			23
VEX	M	2002	1	1			3				3				2			1			11
LISA PF	S	2004					1								1						2
Euclid	M	2012									1										1
JUICE	L	2013		1	1		3				2	2			4	1	2			2	17
CHEOPS	S	2014		3																	3
PLATO	M	2014					1														1
SMILE	S	2015	1								1	3									5
Comet interceptor	S	2019		2		1	1	1	1	2	2		1	1	2	1	1			1	15
ARIEL	M	2020	1		1	1	1	1	2	2	2	2	1	1	2	1	1			1	21
Total			8	16	4	62	5	2	6	8	53	37	1	2	39	8	3	3	1	15	279
% ESA payload			2.87	5.73	1.43	22.22	1.79	0.72	2.15	2.87	19.00	13.26	1.08	0.72	13.98	2.87	1.08	1.08	0.36	5.38
% Fin. contrib.			2.17	3.41	0.92	21.16	1.74	0.12	7.22	1.33	14.94	15.41	0.62	1.03	11.63	4.42	2.16	2.65	1.17	2.58
Ratio			1.32	1.68	1.55	1.05	1.03	6.03	0.30	2.15	1.27	0.86	1.74	0.70	1.20	0.65	0.50	0.41	0.31	2.08
Weighted ratio			0.66	1.08	0.56	1.11	1.62	1.96	0.16	1.19	1.42	0.88	0.99	0.48	1.24	1.85	0.18	0.17	0.14	1.65

payload consortia had been selected by the end of 2021 including PLATO [16] JUICE [17], Comet Interceptor [18] and ARIEL [19]. A total of 422 PIs and co-PIs were identified. We note that the Exosat [20] payload was procured directly by ESA and so is not considered here. The Gaia [13] and Hipparcos payloads were also procured by ESA, but there were substantial national funding contributions to the data processing. These were comparable in scale to the provision of a major instrument and so these missions are included. Both Cluster and GEOS [21] were re-flights, with identical payloads, of earlier missions that failed to achieve their nominal missions—in the case of Cluster due to a launch failure and in the case of GEOS due to a non-nominal final orbit. For each considered mission, a list of instruments was obtained and for each instrument the following information was collected:

1. The names of the PI or PIs and where appropriate any co-PIs
2. The institutes of the PI(s) and any co-PIs
3. The countries of the PI(s) and any co-PIs institutes
4. The gender of the PI(s) and any co-PIs
5. The year of award of PhD degree of the PI(s) and any co-PIs

It should be noted that the consortium that is selected initially often evolves over time, particularly since the mission duration from initial selection to completion may be as long as ~20 years. Changes can occur in consortium composition due to a variety of reasons including retirement of the PI, changing financial, personnel or organisational environment, or other reasons. For this analysis, it was decided to use data pertaining only to the original selections. This left 336 PIs and co-PIs from the ESA-led missions listed in Table 3.1 of which 279 were located in the ESA Member States. This selection represents the thinking of the Advisory Structure and associated infrastructure of ESA and the Member States and thus any biases, conscious or unconscious, are more likely to be contained in those data rather than those resulting from any subsequent evolution of the relevant consortia.

The genders of PIs and any co-PIs were assigned through the personal knowledge of the authors, or the relevant project scientists, or when given on the personal web pages of the individuals concerned, or from other sources (listed below). We appreciate that gender identity is more complex than a binary issue. However, no attempt was made to assign genders other than male or female as this information is not readily available.

In order to determine the "age" or experience of PIs and any co-PIs, we used the year that they obtained their Doctor of Philosophy Degree (PhD) as a proxy. The was determined for the majority of those concerned by searching the internet, particularly sites such as the Astrophysics Data Service (ADS), LinkedIn, the Astronomy Genealogy Project (astrogen.aas.org), ORCID.org, IEEE Xplore (https://ieeexplore.ieee.org/Xplore/home.jsp), and, for French theses https://www.theses.fr. Some proposers were contacted and provided their PhD dates by email. The proposers for which the PhD could not be found are often retired, deceased or have left astronomy. For (co-)PIs who had not yet completed their degrees, the expected year of submission was used. For the small number of proposers who did

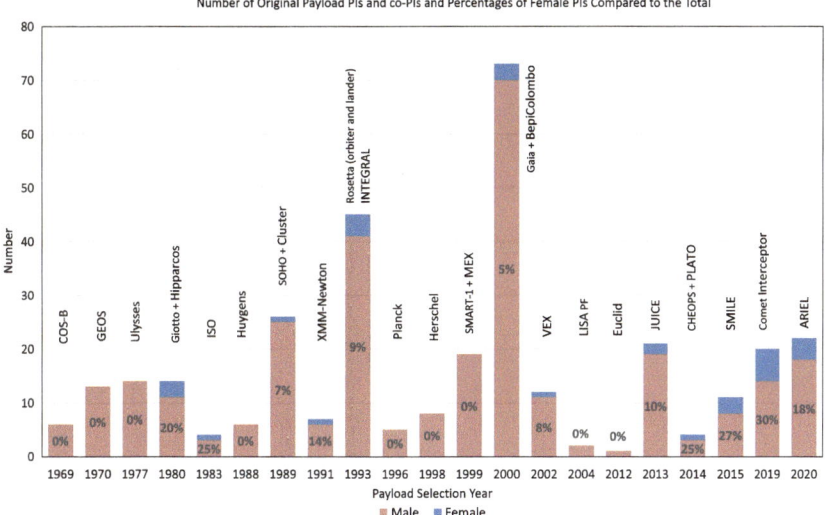

Fig. 3.1 The number of payload PIs and co-PIs for the years in which selections took place. The percentages are the number of female PIs and co-PIs divided by the total number of PIs and co-PIs for each selection year. The increase in the percentage of female PIs and co-PIs from around 2014 to around ∼25% of the total is apparent

not have a PhD and were not enrolled in a PhD programme, their dates were assumed to be arbitrarily far in the future. For some late career scientists in Italian institutes who do not have a PhD, their "age" was taken to be three years after they obtained their Laurea. We note that using the year of PhD to indicate the number of years experience neglects time spent outside of astronomy.

3.3   Payload Contributions

Table 3.1 shows the numbers of original PIs and co-PIs located in the ESA Member States for the ESA-led missions of the Science Programme. These are shown plotted in Figs. 3.1, 3.2 and 3.3. As well as the 279 original PIs and co-PIs from institutes within the ESA Member States, there are an additional 56 PIs and co-PIs located outside the ESA Member States, mainly in the US and Japan with 26 and 15 PIs and co-PIs, respectively. Missions such as Proba 2 [22] and ExoMars [23] where the payloads were not selected by the ESA Science Advisory Structure were excluded from this analysis. All ESA Member States except for Greece, Luxembourg and Romania have PIs or co-PIs located at institutes in their countries with Estonia having its first co-PI in 2019 (Comet Interceptor) and Portugal (ARIEL) in 2020. The four largest financial contributors to the ESA Science Programme (Germany, the United Kingdom, France and Italy) dominate the provision of payloads with

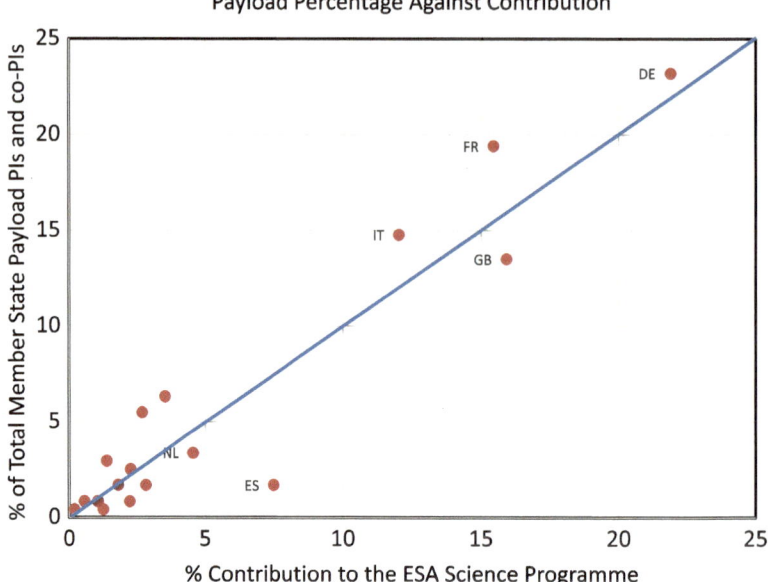

Fig. 3.2 The percentages of PIs and co-PIs providing payload contributions to the ESA-led missions from ESA Member States plotted against their 2000–2021 financial contributions. Member States without PIs or co-PIs are not shown

62, 38, 53 and 39 PIs and co-PIs, respectively. This makes a total of 192 PIs and co-PIs which is 69% of the total located in the ESA Member States. In contrast the same four Member States contributed 63% of the Science Programme funding. The Member States with the next largest number of original PIs and co-PIs are Switzerland and Sweden with 16 and 15, respectively.

The time dependence of the selection of payload PIs and co-PIs is shown in Fig. 3.1. This shows the number of PIs and co-PIs that were selected for the years in which selections took place. The percentages of these PIs and co-PIs that are female is also shown. It is notable that there is generally a very low female participation rate until around 2014 with many of the early selections having no female PIs or co-PIs. From around 2014 $\sim$25% of the selected PIs and co-PIs are female probably reflecting the population of female astronomers measured by the International Astronomical Union (IAU) (see Table 1.2).

The wide range in the number of PIs and co-PIs between different missions is evident in Table 3.1; there are 31 from the ESA Member States on BepiColombo [5] and only one each on PLATO [16] and Euclid [24]. In general, the planetary missions tend to have more PIs and co-PIs than the observatory missions. Table 3.1 shows the number of PIs and co-PIs from each ESA Member State for each mission. These numbers can be compared to the average relative contributions to the ESA Science Programme funding between 2000 and 2021 which we take to be the expected relative payload contribution. The ratio of these two numbers gives an

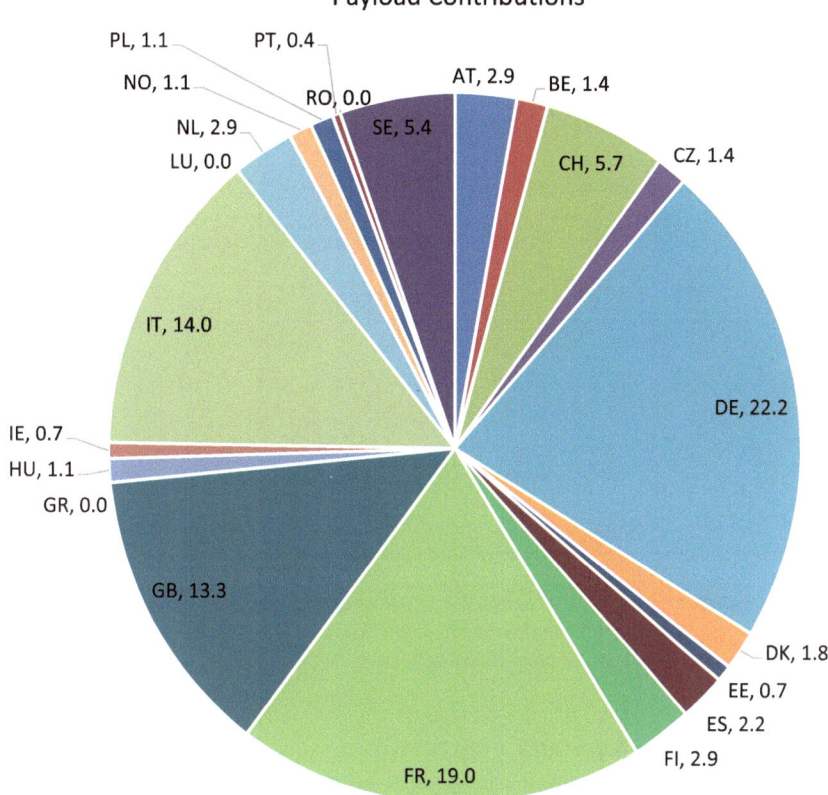

Fig. 3.3 The percentages of PIs and co-PIs providing payload contributions to the ESA-led missions from ESA Member States. Member States without PIs or co-PIs are not shown

indication of whether a particular ESA Member State has provided more or less payload elements than expected. These ratios are listed in the bottom section of Table 3.1 and plotted against the percentage contribution in Fig. 3.4.

Examining Fig. 3.4 and Table 3.1 shows that of the "big four" contributors, who between them provide 63% of the Science Programme funding cluster together on the figure. It can be seen that Germany has a payload contribution comparable to its financial contribution, The United Kingdom provides less payload contribution than expected and both Italy and France provide more payload contributions than expected. The fifth biggest funding contributor is Spain which has a very low payload contribution, compared to expectation. It is more difficult to draw conclusions for some of the smaller countries due to the small number of original PIs and co-PIs involved.

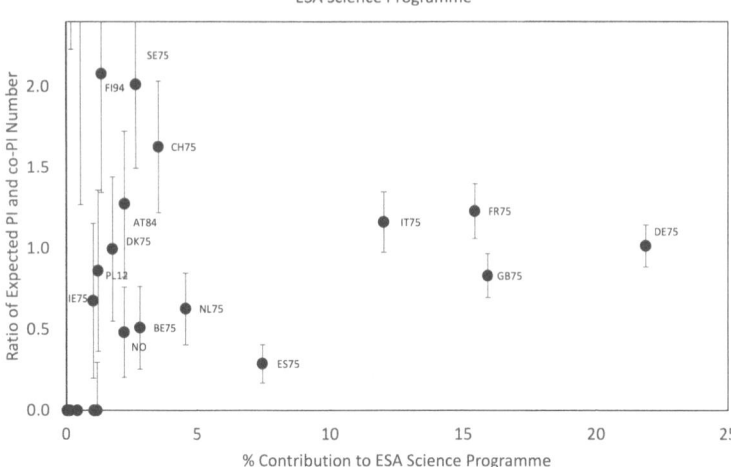

Fig. 3.4 The ratio of the number of PIs and co-PIs divided by the total number from the ESA Member States normalised by the contributions of each Member State to the funding for the Science Programme plotted against the contributions. The last two digits of the year in which each country joined ESA is indicated in its label. The three Member States that have no PIs or co-PIs (Greece, Luxembourg, and Romania) are not included. The uncertainties shown are indicative only as they assume the PI and co-PI numbers follow Poisson statistics

3.4 Payload Complexity and Size

We are aware that simply counting the number of PIs and co-PIs may bias the results towards Member States that support relatively more planetary missions compared to astronomy missions due to the larger number of payload elements on planetary missions. For that reason we have recalculated the ratios of original PI and co-PI numbers compared to the financial contributors by first applying a series of weighting factors. We assigned an overall weighting factor to each mission. This is the product of three different factors:

1. The mission class
2. The number of worldwide PIs and co-PIs for a mission
3. The relative contribution of the average co-PI compared to a PI

3.4.1 Mission Class

A complicating factor in any such comparative study is the fact that space science missions have different complexities, sizes and therefore costs. In order to investigate whether such differences could affect the analysis of the payload contributions presented above, all of the missions under consideration have been ascribed a weighting factor. The procedure that was chosen mirrors the broad system of mission classification which has been used within the ESA Science Programme in recent years. The ESA Science Programme has been mostly centred around so-called M, L & S class missions where these designations stand for Large, Medium and Smart/Small class missions. In order to reflect the differences in magnitude of such missions, we have used weighting factors of 2, 1 and 0.3 to reflect likely differences in payload costs which are broadly assumed to scale with the cost to ESA of each mission class. The designation for each mission is shown in Column 2 of Table 3.1 (Class).

3.4.2   Number of PIs and co-PIs

As discussed earlier, different missions can have very different numbers of instruments, varying from one or very few for observatory missions up to ten or more for typical planetary science missions. In order to account for the very different number of instruments as well as the different sizes of mission, we have divided the mission class weighting factors by the number of worldwide PIs and co-PIs (counted as half a PI, see below) before calculating the contributions of each country. The number of PIs and co-PIs for each mission is shown in the right-most column of Table 3.1.

3.4.3   Co-PI Contributions

Most instruments have a single PI but sometimes there are a number of co-PIs. These generally reflect a funding contribution to the payload from the co-PI's nation. For the purpose of this study we have assumed that on average co-PIs contribute payload with half the value of that of a PI. We therefore included the number of PIs and 0.5 times the number of co-PIs from a particular Member State as a weighted measure of its contribution to the ESA Science Directorate missions' payloads. Figures 3.5 and 3.6 show the weighted contributions for the relevant Member States.

3.5   Effect of Weighting Factors

We then recalculated the ratios of expected payload contributions as before using the three factors described above. These are shown in the bottom line of Table 3.1 "Weighted Ratio" and are plotted against ESA Member State funding contributions in Fig. 3.7.

Comparison of Figs. 3.4 and 3.7 and the ratios given in Table 3.1 shows that the application of weighting factors has very little effect on the outcomes for the

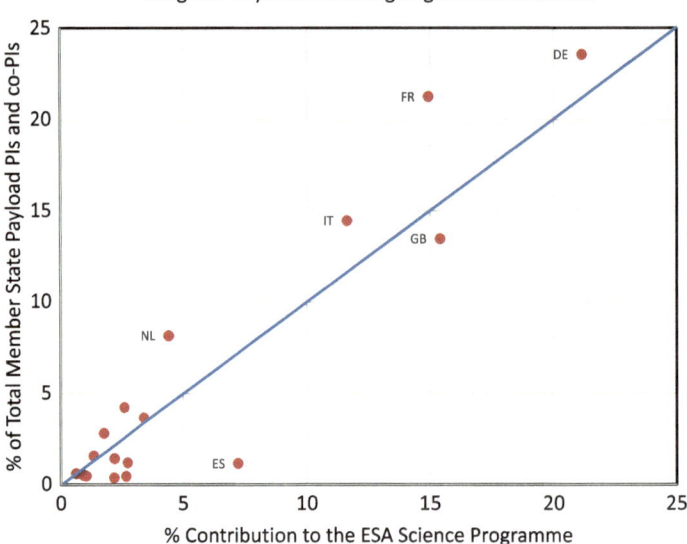

Fig. 3.5 The weighted percentages of PI and co-PIs providing payload contributions to the ESA-led missions from ESA Member States plotted against their 2000–2021 financial contributions to the Programme. Member States without payload PIs or co-PIs are not shown

four large ESA Member States with the ratios increasing by 0.06, 0.15 and 0.04 for Germany, France and Italy, respectively. The ratio remained unchanged for the United Kingdom. These small changes indicate a robust process and that it is not strongly affected by payload complexity for Member States with $\gtrsim 30$ PIs and co-PIs. For the next largest contributor to the ESA Science Directorate funding (Spain) there is a small decrease in the ratio when weighting is applied of 0.14 while the ratio increases strongly for the Netherlands from 0.65 to 1.85 when weighting is applied. This is likely the result of strong Dutch provision of observatory payloads with their typically smaller number of instruments. For the other ESA Member States it is difficult to draw reliable conclusions from the application of weighting factors due to the smaller number of PIs and co-PIs involved. However, comparison of Figs. 3.4 and 3.7 shows that the application of weighting factors tend to reduce the relative contributions of the smaller Member States.

3.6 Payload PI and co-PI Genders

We have examined the number of original female PIs and co-PIs compared to the total. Of a total of 336 such PIs (from Member and non-Member States) 32 were female. which is 9.5% of the total. This is much less than the fractions of female observing time PIs on the observatory missions INTEGRAL, Herschel and

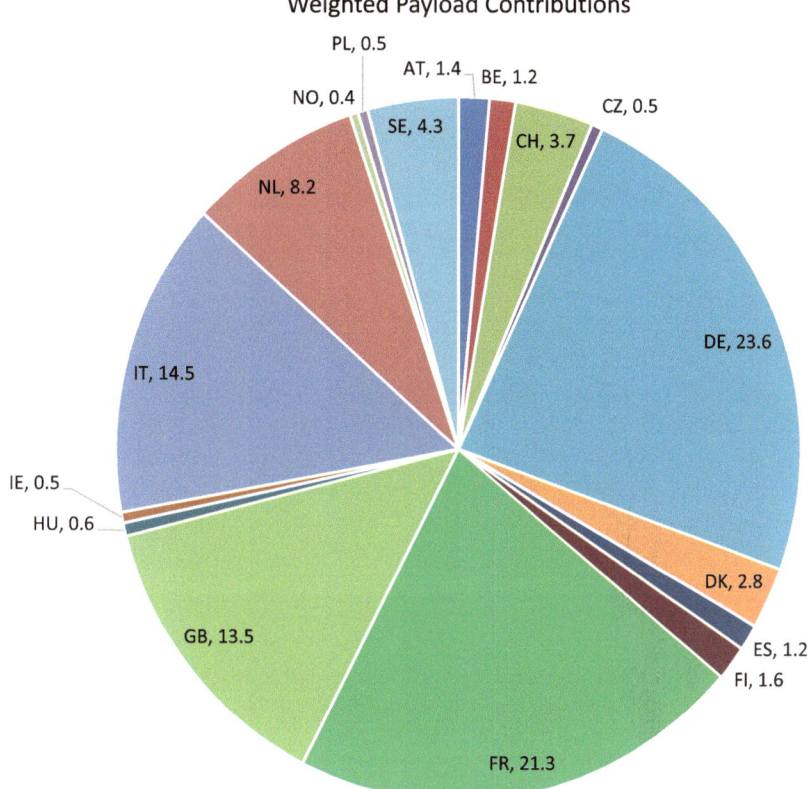

Fig. 3.6 The weighted percentages of PI and co-PIs providing payload contributions to the ESA-led missions from ESA Member States. Member States without PIs or co-PIs are not shown. The method of calculating the weighting factors is described in Sect. 3.4

XMM-Newton. The clearest example is with XMM-Newton where the fraction of proposals from female PIs increases from ∼20% to ∼30% of the total over the 21 years of the mission that is being considered (see Fig. 4.8). This is broadly comparable with the gender distribution of the IAU membership which ranges from ∼50% female between the ages of 25–30 to ∼18% for ages 60–65 (see Table 1.2).

Being a PI or co-PI on an instrument can be considered to be a leadership position in the community, so these are likely to be senior people with substantial experience. To check this, we determined the average difference between the year that the payload of a mission was approved and the year in which each originally selected PIs and co-PIs received their Doctor of Philosophy Degree (PhD)s. We searched the internet, particularly sites such as the ADS, LinkedIn, the Astronomy Genealogy Project (astrogen.aas.org) and ORCID.org. For Italian astronomers without a PhD we used the year the Linea was obtained plus three years. We found that for the 84% of the 337 PIs and co-PIs where we found the year they obtained their PhD,

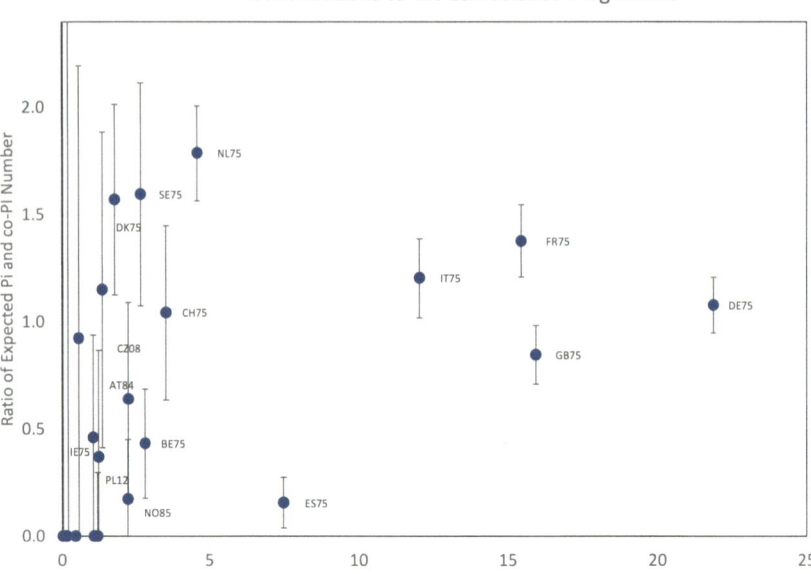

Fig. 3.7 The weighted ratio of the number of PIs and co-PIs divided by the total number from the ESA Member States divided by the contributions of each Member State to the funding for the Science Programme plotted against the contributions. The year in which each country joined ESA is indicated in its label. The three Member States that have not provided payload elements (Greece, Luxembourg and Romania) are not included

the average difference is 15.9 years, consistent with a more "senior" population. For comparison, the average "academic ages" of XMM-Newton, INTEGRAL and Herschel observing time PIs are 10.9, 13.7 and 12.2 years, respectively (see Sects. 4.8.3, 5.8.4 and 6.9). For male payload PIs the average is 15.7 years and for female PIs 17.9 years. If we assume an average age for obtaining a PhD of 25–30 years, this implies an average PI and co-PI age of 40–45 years. The current IAU membership for this age range comprises of around 26% females (see Table 1.2). This is similar to the percentage of female PIs and co-PIs in recent payload selections (~25%) suggesting that the chance of selection for male and female PIs is comparable.

3.7 Conclusions

The number of payload PIs and co-PIs has been used as a proxy of the contributions of the different ESA Member States to the payloads of the ESA Science Directorate's missions. The four largest financial contributors to the ESA Science Programme (Germany, the United Kingdom, France and Italy) dominate the provision of payloads with in total 192 original PIs and co-PIs which is 68% of the total located in all ESA Member States. It is pleasing to see that the percentage of female PIs and co-PIs has increased in recent selections to reach ∼25%. This is comparable to the current population of "senior" female IAU members.

In order to continue studies similar to this, and to monitor any future trends, it would be invaluable if ESA and the ESA Member States would:

1. Encourage prospective instrument teams to include a breakdown of the gender distribution of their members, including the PI. This would allow much easier monitoring of the gender outcomes of the selection processes.
2. Take due account, when instrument team proposals are assessed, of the gender of the PI and the gender distribution of the members. We expect that the number of female instrument team PIs will increase with time given the increasing fraction of early career female astronomers (see Table 1.2) who are now beginning to reach "senior" positions. We note that recent payload PI selections have resulted in ∼25% of female PIs, which is significantly higher than earlier selections. Nevertheless, scientific excellence should always remain the main driver for selection.
3. Establish a mentoring system for new and prospective instrument PIs which would help guide less senior scientists in this sometimes daunting task.
4. Establish an open database containing all information, in a homogeneous form, pertaining to payload teams, including institutions, academic ages, gender and any other information to enable these factors to be monitored and updated on a regular basis.

Acknowledgments We thank Erik Kuulkers for helping find the PhD dates for many of the PIs and co-PIs and Peter Wenzel, Arnaud Masson, Cecil Tranquille, Steve Sembey, Michael Küppers, Philippe Escoubet, Håkan Svedhem, Matt Taylor, Johannes Benkhof and Colin Wilson for help in finding the original payload PIs and co-PIs. This research was supported by the International Space Science Institute (ISSI) in Bern, through the ISSI Working Group project "The Scientific Performance of ESA's Science Missions".

References

1. J.P. Lebreton, D.L. Matson, The huygens probe: science, payload and mission overview. Space Sci. Rev. **104**(1), 59–100 (2002)
2. G. Israel, H. Niemann, F. Raulin, et al., The aerosol collector pyrolyser (ACP) experiment for Huygens. ESA SP-1177 (1997), pp. 59–84
3. H. Niemann, et al., The gas chromatograph mass spectrometer aboard huygens. ESA SP-1177 (1997), pp. 85–107
4. H. Svedhem, D. Titov, F. Taylor, et al., Venus express mission. J. Geophys. Res. **114**, E00B33 (2009)
5. J. Benkhoff, G. Murakami, W. Baumjohann, et al., BepiColombo - mission overview and science goals. Space Sci. Rev. **217**(8), 90 (2021)
6. V. Domingo, B. Fleck & A.I. Poland, The SOHO Mission: An Overview. Solar Phys. **162**(1), 1–37 (1995)

7. R. Schulz, C. Alexander, H. Boehnhardt, K. Glassmeier, Rosetta: ESA's Mission to the Origin of the Solar System. Springer (2009)
8. R. Reinhard, The Giotto mission to Halley's comet. Adv. Space Res. **2**(12), 97–107 (1982)
9. R. Marsden, B. Fleck, The Solar Orbiter mission, in *34th COSPAR Scientific Assembly*, vol. 34 (2002), p. 222
10. E. Kuulkers, C. Ferrigno, P. Kretschmar, et al., INTEGRAL reloaded: spacecraft, instruments and ground system. New Astron. Rev. **93**, 101629 (2021)
11. F. Jansen, D. Lumb, B. Altieri, et al., XMM-Newton observatory. I. The spacecraft and operations. Astron. Astrophys. **365**, L1–L6 (2001)
12. G.L. Pilbratt, J.R. Riedinger, T. Passvogel, et al., Herschel space observatory. An ESA facility for far-infrared and submillimetre astronomy. Astron. Astrophys. **518**, L1 (2010)
13. M.A.C. Perryman, The GAIA mission, in *GAIA Spectroscopy: Science and Technology*, ed. by U. Munari. Astronomical Society of the Pacific Conference Series, vol. 298 (Society of the Pacific Astronomical, San Fransisco, 2003), p. 3
14. A. Piccialli, J.A. Rathbun, A.C. Levasseur-Regourd, et al., Participation of women scientists in ESA solar system missions: a historical trend. Adv. Geosci. **53**, 169–182 (2020)
15. L. Scarsi, K. Bennett, G.F. Bignami, et al., The Cos-B experiment and mission, in *Recent Advances in Gamma-Ray Astronomy*, ed. by R.D. Wills, B. Battrick, vol. 124 (ESA Special Publication, Paris, 1977), p. 3
16. H. Rauer, C. Catala, C. Aerts, et al., The PLATO 2.0 mission. Exp. Astron. **38**(1–2), 249–330 (2014)
17. O. Grasset, D. Titov, JUICE mission, in *Encyclopedia of Astrobiology*, ed. by M. Gargaud, W.M. Irvine, R. Amils, et al. (Springer, Berlin, 2015), pp. 1309–1313
18. G. Jones, C. Snodgrass, C. Tubiana, The comet interceptor mission, in *European Planetary Science Congress* EPSC2022-1085 (2022)
19. G. Tinetti, P. Drossart, P. Eccleston, et al., A chemical survey of exoplanets with ARIEL. Exp. Astron. **46**(1), 135–209 (2018)
20. B.G. Taylor, R.D. Andresen, A. Peacock, et al., The EXOSAT mission. Space Sci. Rev. **30**(1–4), 479–494 (1981)
21. K. Knott, The GEOS-1 mission. Space Sci. Rev. **22**, 321–325 (1978)
22. S. Santandrea, K. Gantois, K. Strauch, et al., PROBA2: mission and spacecraft overview. Solar Phys. **286**(1), 5–19 (2013)
23. J.L. Vago, B. Gardini, P. Baglioni, et al., Science objectives of ESA's ExoMars mission, in *European Planetary Science Congress 2006* (2006), p. 76
24. R.J. Laureijs, L. Duvet, I. Escudero Sanz, et al., The euclid mission, in *Space Telescopes and Instrumentation 2010: Optical, Infrared, and Millimeter Wave*, ed. by J.M. Oschmann, Jr., M.C. Clampin, H.A. MacEwen, vol. 7731. Society of Photo-Optical Instrumentation Engineers (SPIE) Conference Series (2010). 77311H

Chapter 4
XMM-Newton Observing Time Proposals

Arvind Parmar, Norbert Schartel, and Maria Santos-Lleó

Abstract We examine the outcomes of the regular announcements of observing opportunities for ESA's X-ray observatory XMM-Newton issued between 2001 and 2021. We investigate how success rates vary with the lead proposer's gender, "academic age" and the country where the proposer's institute is located. The large number of proposals (10,579) and more than 20 years operational lifetime enable the evolution of community proposing for XMM-Newton to be examined. We determine proposal success rates for high-priority and all proposals using both the numbers of accepted proposals and the amounts of awarded observing time. We find that male lead proposers are between 5–15% more successful than their female counterparts in obtaining XMM-Newton observations. The gender balance and the percentage of successful young proposers are comparable to those of HST after the introduction of dual-anonymous reviewing of HST proposals. We investigate potential correlations between the female-led proposal success rates and the amount of female participation in the Time Allocation Committee. We propose additional investigations to better understand the outcomes presented here.

Keywords XMM-Newton · Science results · Science productivity · Users' Group · Announcement of observing opportunities · Over-subscription · Open time · Peer review · Time allocation committee · Panel members · Panel chairs · Observing priorities · Gender · Gender balance · Proposers' institutes countries · Academic age · Proposal acceptance rates · Regional dependence · Dual-anonymous reviews

A. Parmar (✉)
Department of Space and Climate Physics, MSSL/UCL, Dorking, UK
e-mail: arvind.parmar@ucl.ac.uk

N. Schartel · M. Santos-Lleó
Directorate of Science, ESA, ESAC, Villanueva de la Cañada, Madrid, Spain
e-mail: norbert.schartel@esa.int; maria.santos-lleo@esa.int

© The Author(s) 2025
A. Parmar et al., *ESA Science Programme Missions*, ISSI Scientific Report Series 18,
https://doi.org/10.1007/978-3-031-69004-4_4

4.1 Introduction

XMM-Newton was launched on 10 December 1999 into a 48-hour highly elliptical orbit. The nominal mission lifetime was 5 years with a designed lifetime of 10 years. The mission provides sensitive X-ray imaging and spectroscopic observations of a wide variety of cosmic sources from nearby solar system objects to the most distant black holes [1]. The payload consists of the European Photon Imaging Camera (EPIC); [2], Reflection Grating Spectrometer (RGS); [3]) and the Optical Monitor (OM); [4]. The EPIC consists of three imaging spectrometers each located at the focus of an X-ray optic consisting of 58 nested Wolter I geometry mirrors. Two of the EPIC cameras are based on MOS-CCD technology and share the mirrors with RGS grating arrays [5] while the detector based on pn-CCD technology is located behind a fully open telescope [2]. The overall effective area is 2500 cm^2 at 1.5 Kilo Electron Volt (keV) and the spatial resolution is 15 arc seconds (half-energy width) and 6 arc seconds (Full-Width at Half Maximum (FWHM)) with a Field of View (FOV) of $\sim$30 arc minutes diameter. The RGS provides 0.35–2.4 keV spectra with an E/ΔE of 300–700 (1st order). The effective area for the two grating arrays varies in the range of 40–110 cm^2 over the energy range. The OM provides optical and Ultra-Violet (UV) monitoring of fluxes through various filters as well as spectroscopy with two grisms. Normally, all three instruments are operated simultaneously.

The scientific products from each observation, produced by a standard pipeline processing, are placed into the XMM-Newton Science Archive (XSA). During any proprietary period data in the XSA are only made available to the Principal Investigator (PI) of the observation, otherwise all the products stored in the XSA are available for public download via the internet. More than 2000 users download data each month. Investigators located in the USA are served from the Goddard Space Flight Center (GSFC) archive which also provides proprietary data for US PIs. General information such as observing date and time, instrument use, the name of PIs etc. can be accessed by the general public for all observations in the XSA. Also, links to the refereed papers that make use of data from each observation can be obtained from the XSA. The key to accessing a data set belonging to a specific observation is the unique 10-digit Observation Identifier (ObsID), consisting of the six-digit proposal number and the four-digit observation identifier.

4.2 Science Results

XMM-Newton is a highly successful observatory with over 6900 refereed publications by the end of 2021. It was launched shortly after National Aeronautics and Space Administration (NASA)'s Chandra X-ray observatory [6]. The two missions are highly complementary with Chandra providing higher spatial resolution imaging whilst XMM-Newton has a larger collecting area. XMM-Newton observes celestial sources ranging from charge-exchange radiation in the upper Earth's magnetopause [7] to the most distant quasars [8–11]. Observations with these two missions have changed X-ray astronomy from being a specialised pursuit to one that contributes to nearly all areas of modern observational astrophysics. The scientific impact of XMM-Newton and Chandra is so far reaching that Nature published reviews of their impact in 2009 [12] and again in 2022 [13]. In the following we give some examples that illustrate the scientific impact of XMM-Newton:

Solar System XMM-Newton observations have lead to the detection of charge exchange radiation from Mars [14]. Charge exchange occurs when a neutral atom and an ion interact and an electron is transferred together with the emission of one or more photons. This was the fist time that this mechanism was identified in the radiation from another planet. XMM-Newton and Chandra found independent pulsations of Jupiter's northern and southern X-ray aurorae [15]. Simultaneous observations by XMM-Newton and the Juno [16] spacecraft revealed the source of Jupiter's auroral flares [17] which are explained through a pulsating magnetic field caused by interactions with the solar wind.

Exoplanets XMM-Newton and Chandra observations discovered the first planet in another galaxy [18]. The study of exoplanets and the interactions with their host stars is an emerging area of X-ray astrophysics. XMM-Newton observation have revealed that the corona of the star HD 189733 flares in phase with the orbit of its hot-Jupiter exoplanet [19]. Helium absorption is an important diagnostic of exoplanet atmospheres, particularly in hot Neptune and Jupiter systems which may have thick evapourating atmospheres producing gaseous tails streaming behind the planet. XMM-Newton observations have revealed the dependence of the Helium absorption on a sample of host stars' X-ray luminosities [20] allowing an improved characterisation of the atmospheres and contributing physical processes to be made.

Millisecond Pulsars XMM-Newton observations were key in identifying a millisecond pulsar which cycles between emitting radio and X-rays. The object is the evolutionary link between accretion and rotation-powered millisecond pulsars. The existence of such objects had been predicted, but never found [21]. Soft Gamma-ray Repeaters are slowly rotating isolated neutron stars that sporadically undergo outbursts. These are assumed to be powered by their internal magnetic energy with dipole field strengths of 10^{14}—10^{15} Gauss and are often called "magnetars". A surprise was the detection of a Soft Gamma-ray Repeater, SGR 0418+5729, with a low-magnetic-field strength as determined from its timing properties [22]. XMM-Newton detected a rotation-phase dependent proton cyclotron absorption feature from this source, which can be interpreted as originating from a multi-pole component of the magnetic field which has a strength in the range of expectation for magnetars [23]. An ultra-luminous X-ray source is an exceptionally bright X-ray source that is not coincident with its galactic nucleus. The detection of pulsations from some of these sources indicates a neutron star origin [24, 25] and therefore excludes intermediate-mass black holes for at least part of the population.

Supermassive Black Holes XMM-Newton has been used to investigate the environments close to supermassive black holes using a technique called reverberation mapping. This utilises variations in the X-ray continuum emission from the corona and the X-ray line emission originating in the inner regions of the accretion disk which "reverberate" with a time delay due to the light travel time [26], see also [27]. The current state of the art was demonstrated by XMM-Newton observations of the highly variable Active Galactic Nucleus (AGN) IRAS 13244-3809 [28]. The XMM-Newton campaign allowed the inherent degeneracy between the black hole mass, inner disk radius and height of the X-ray emitting corona to be broken. This allowed, for the first time, the simultaneous determination of the supermassive black hole mass and spin. A flare during the observation of the AGN 1 Zwicky 1 allowed the light bending and X-ray echos from behind a supermassive black hole [29] to be traced. XMM-Newton was of fundamental importance in the detection and exploration of fast and ultra-fast outflows of highly ionized matter from the nuclei of AGNs [30–34]. The growth of galaxies and cluster of galaxies is significantly slower over cosmological time scales than predicted by simulation. Outflows from AGNs are a strong candidate explain this lower growth through feedback on the cosmological structure formation.

Tidal Disruption Events XMM-Newton has been used to study the Tidal Disruption Event (TDE)s of stars captured by black holes. This allows the properties of black holes in the centre of normal galaxies to be investigated [35]. Combining observations from XMM-Newton and Chandra a TDE from an elusive intermediate-mass black hole in a star cluster was detected [36] and a decade long TDE was traced [37]. Quasi-periodic oscillations were detected from two TDEs. These probably originate from areas close to the event horizons of the black holes [38, 39]. Blue-shifted absorption lines of highly ionized atoms may also be explained though emission close to the black hole's event horizon [40]. Even relativistic reverberation in the accretion flow of one TDE was evident based on XMM-Newton data [41]. XMM-Newton leads in the detection and exploration of quasi-periodic eruptions[42, 43], which are most likely explained by localised disruptions of accretion disks around supermassive black holes by orbiting compact objects following (partial) tidal disruption events.

Cluster of Galaxies XMM-Newton observations have allowed a universal mass profile for clusters of galaxies to be established [44] as well as determining the origin of the metals observed in them [45, 46]. Some 30% of the metals can be traced to Type I supernovae and 70% to core collapse supernovae. One of the fist achievement of XMM-Newton was the non-detection of strong X-ray emission from cooling flows in the centre of clusters of galaxies as was generally expected [47–49]. The low upper-limits on turbulence induced broadening of emission lines in the XMM-Newton RGS spectra of some of the brightest clusters of galaxies, indicates the dissipation of turbulence may prevent cooling of the cluster core [50, 51].

Cosmology An RGS observation of a distant AGN has allowed the detection of the Warm-Hot Intergalactic Medium (WHIM) for the first time [52] closing the gap between the number of observed baryons and number of baryons predicted by big-bang nucleosynthesis. By combining gravitational lensing observations of the total mass with optical and infrared observations of the cold baryonic mass and XMM-Newton observations of the hot baryonic mass in the WHIM, the dark matter large-scale distribution in the Cosmos field has been determined [53]. It is found to be consistent with the predictions of cosmic structure formation. XMM-Newton plays an important role in the search for dark matter particles [54–56]. An analysis of $\sim$450 clusters observed by XMM-Newton in $\sim$50 square degrees of the extra-galactic sky shows clear tensions with the predictions using the cosmological parameters as determined by Planck 2015 [57, 58]. An analysis of the XMM-Newton follow-up observations of Planck detected clusters of galaxies [59] also reveals tensions with the predictions using the standard concordance cosmological model, as does an analysis of the Hubble Diagram for quasars [60].

4.3 Scientific Productivity

The scientific productivity of XMM-Newton was assessed in 2014 by Ness et al. [61]. For comparison, similar analyses for NASA's Chandra X-ray observatory are to be found in [62] and for Hubble Space Telescope (HST) in [63] as well as for a range of astronomical facilities in [64–67]. The ObsID and instruments used to provide the scientific results reported in each publication were determined by Ness et al. [61]. The ObsIDs were then used to access the XSA to provide detailed information on the observations themselves and on the original proposal. The information obtained from these sources was then combined to allow the scientific productivity of the mission to be investigated. The major conclusions of [61] were:

- Around 100 scientists per year become lead authors for the first time on a refereed paper which directly uses XMM-Newton data.
- Each refereed XMM-Newton publication receives an average of around four citations per year with a long-term citation rate of three citations per year, more than five years after publication.
- About half of the articles citing XMM-Newton articles are not primarily X-ray observational papers.
- The distribution of elapsed time between observations taken under the Guest Observer program and first article peaks at 2 years with a possible second peak at 3.25 years. Observations taken under the Target of Opportunity (ToO) program (see below) are published significantly faster, after one year on average.
- The fraction of science time used in at least one publication reaches >95% after 7 years and about 70% after exclusion of catalog-type publications.
- The scientific productivity of XMM-Newton measured by the publication rate, number of new authors and citation rate, remained extremely high with no evidence that it was decreasing after more than (the then) 12 years of operations.

Up to the end of 2021, there have been over 6900 XMM-Newton refereed publications that meet the inclusion criteria given in Sect. 2.2.

4.4   Observing Time

The annual calls for observing proposals are for "Open Time" targets. There are four types of XMM-Newton observing time:

1. Open Time: ~90% of the observing time is available for Guest Observers via Announcement of Opportunity (AO)s which are open to scientists worldwide. Open Time includes anticipated ToOs – astronomical events observable by XMM-Newton, which cannot be predicted and scheduled at the time of proposal submission.
2. Discretionary Time: The XMM-Newton Project Scientist can grant "Discretionary Time" which is ~5% of the available observing time. This can be used for rapid unanticipated ToOs – observations where a timely observation outside the normal AO cycle is likely to result in a significant scientific impact. Discretionary Time observations may have a proprietary period of six months, or may be made publicly available as soon as the relevant data files have been created.
3. Calibration Time: ~5% of the available observing time is used to maintain the calibration and monitor instrument health.
4. Guaranteed Time: During the first two years of the mission, the scientific groups that provided the instruments and that were involved in the scientific ground segment development were awarded Guaranteed Time observations.

Open Time targets are selected competitively through peer review by an Observation Time Allocation Committee (OTAC). Calls for observing proposals are usually issued annually and are highly oversubscribed (by at least a factor of 5) compared to the available observing time (see Fig. 4.1). Proposals are submitted electronically to the XMM-Newton Science Operations Centre (SOC). Following receipt proposals are submitted to the OTAC for scientific review while SOC staff perform visibility assessments and duplication checks. The feasibility assessment is performed by panel members of the OTAC. The PIs of successful proposals are required to provide detailed information on the requested configurations in a second submission phase. The PIs of successful proposals are granted a proprietary period of one year before the data are made publicly available. The types of Open Time proposals available has evolved as the mission's science has matured and as new opportunities to collaborate have become available. For AO-20, these included:

1. Guest Observer. These are "normal" observations including time constrained and observations coordinated with other facilities.
2. Anticipated ToOs.
3. Large Programmes requiring a significant amount of observing time (>300 ksec).

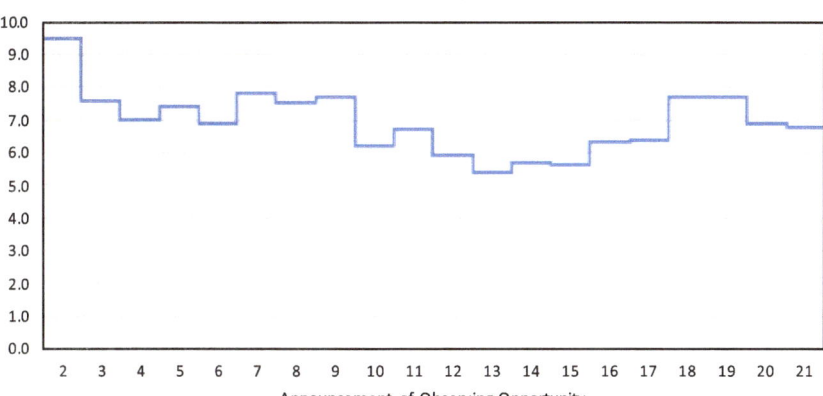

Fig. 4.1 The over-subscription of requested XMM-Newton observing time compared to that available for AO-2 to AO-21, covering an interval of 20 years. The values for AO-17 and AO-20 exclude the multi-year heritage programme opportunities which were offered during these AOs

4. Multi-Year Heritage Programmes. These are scientific visionary programmes which need between 2 and 6 Msec of observing time and are performed over three AOs.
5. Fulfil Programmes: To assist in the completion of samples of objects.
6. Joint observations with INTEGRAL, Chandra, and the European Southern Observatory (ESO) Very Large Telescope (VLT), HST, Swift, NuSTAR, High Energy Stereoscopic System (HESS), MAGIC and NRAO/GBO observatories.

With the introduction of Multi-Year Heritage Programmes in 2017, up to four primary investigators are allowed for large programs (>300 ksec) in order to promote cooperation between diverse groups and to support raising the funding necessary for extended data analysis. In these cases, the corresponding author is assumed to be the PI.

4.5   Observation Time Allocation Committee (OTAC)

The OTAC reviews all submitted proposals and makes recommendations on the targets to be observed by XMM-Newton. It takes into account the scientific case, justification, merit and relevance of the proposed observation(s) and the potential contribution of the overall scientific return of the mission. In addition, potential duplication with performed and planned observations with XMM-Newton and Chandra, as well as the technical feasibility, are also considered. The OTAC recommendations on the observing programme are given priority assignments of A, B and C.

Table 4.1 XMM-Newton AO summary. The numbers of OTAC members include panel chairs

AO	Year issued	No. of proposals	Over-subscription (time)	OTAC Members		OTAC Panel Chairs	
				Male	Female	Male	Female
2	2001	869	9.5	41	6	13	2
3	2003	692	7.6	59	11	12	2
4	2004	660	7.0	58	6	12	1
5	2005	641	7.4	56	9	13	0
6	2006	594	6.9	56	9	13	0
7	2007	586	7.8	59	6	12	1
8	2008	555	7.5	52	13	10	3
9	2009	539	7.7	53	12	12	1
10	2010	491	6.2	53	17	13	1
11	2011	501	6.7	57	13	12	2
12	2012	475	5.9	53	12	12	1
13	2013	452	5.4	50	15	11	2
14	2014	431	5.7	53	12	12	1
15	2015	431	5.6	53	12	10	3
16	2016	442	6.3	46	19	10	3
17	2017	441	6.4	51	15	8	5
18	2018	442	7.7	48	14	7	6
19	2019	462	7.7	54	11	8	5
20	2020	447	6.9	47	16	7	5
21	2021	428	6.8	33	22	6	5
Total		**10,579**		**1032**	**250**	**213**	**49**

- Priority A – The highest priority targets (about 10% of the available observing time) that are of major scientific importance and are scheduled with the highest priority. Observations that are not completed within the observing period are automatically transferred to the next one.
- Priority B – As per Priority A, but for the next highest 40% of observing time.
- Priority C – Targets that are used as "fillers" for around 50% of the observing time and have a significantly lower chance of being observed. Targets that have not been observed by the end of the observing period are not transferred to the next one.

An average of 530 valid proposals were received in response to each AO (see Table 4.1). With an over-subscription factor that remains high during all the AOs (see Fig. 4.1 and Table 4.1). The scientific assessment of the proposals is a major undertaking by the astronomical community with a significant fraction of the high-energy community having participated so far. The OTAC is divided into different scientific categories to reflect the range of topics proposed. These categories have changed over time reflecting the scientific evolution and consequently different number of requests in the various categories. Currently the categories include stars and planets, compact objects, active galactic nuclei, individual, groups and cluster of galaxies, and cosmology. Much of the assessment is performed by panels, consisting of five scientists selected from the worldwide community. The majority of OTAC members are selected from the participants of previous AOs, i.e. scientists who proposed for XMM-Newton time are often invited to be members of subsequent OTACs. Many categories require multiple panels and there are typically between 11 and 15 panels for each AO including dedicated panels for Multi-year Heritage Programmes. Each panel is led by a panel chair and the overall OTAC is led by the chairperson. The names of the OTAC members and chairs (with exception of the chairperson who sits on the XMM-Newton Users' Group (XUG), see Sect. 4.6) are not public.

The genders of the OTAC members and panel chairs were assigned through the personal knowledge of the authors and SOC staff. We appreciate that gender identity is more complex than a binary issue. However, no attempt was made to assign genders other than male or female in this study. We then examined the gender composition of the OTAC members and panel chairs (Table 4.1) over the 20 years between AO-2 to AO-21. AO-1 was opened prior to the XMM-Newton launch and used different software which is no longer maintained. This means that some of the information needed for comparisons with later AOs is not available. For this reason only the results of AO-2 to AO-21 are considered here.

OTAC panel members are chosen for their high-level of relevant scientific knowledge, while selection of the OTAC chairs is for scientists considered to have leadership roles in high-energy astronomy. Figure 4.2 shows the fraction of female XMM-Newton OTAC members from AO-2 to AO-21 and Fig. 4.3 the same for the panel chairs. In total there were 1032 male and 250 female panel members and 213 male and 49 female panel chairs. Since members and chairs may serve for multiple

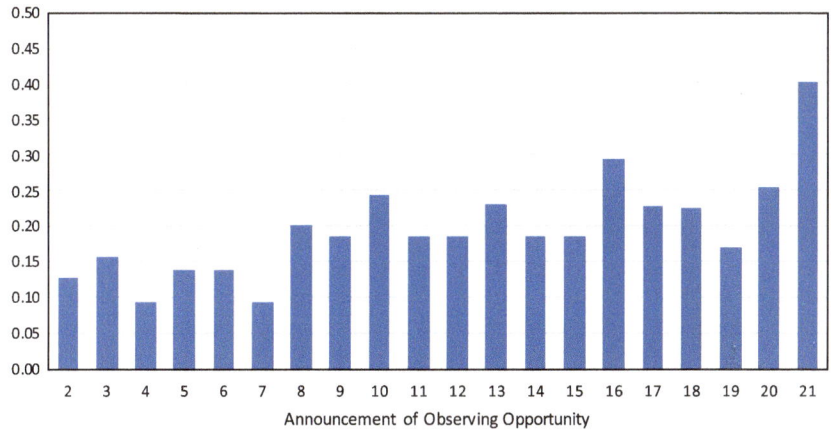

Fig. 4.2 The fraction of female XMM-Newton OTAC members compared to the total for AO-2 to AO-21 covering an interval of around 20 years. A steady increase in the fraction of females from ~0.15 to ~0.25 by AO-20 is evident with a sharp increase in AO-21

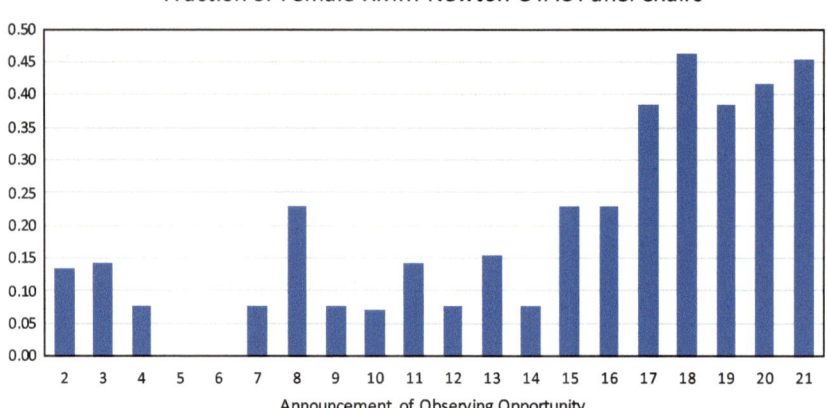

Fig. 4.3 The fraction of female XMM-Newton OTAC chairs compared to the total for AO-2 to AO-21 covering an interval of around 20 years. An increase in the fraction of females from ∼0.10 to ∼0.45 by AO-21 is evident

AOs, the numbers of individuals involved is smaller. Most of the OTAC members (and chairs) are based at institutes located in the European Space Agency (ESA) Member States with a significant number from institutes in the United States. An overall increase in the fraction of females from ∼0.15 to ∼0.25 by AO-20 is evident with a sharp increase in AO-21. There is a marked increase in the fraction of females in leadership positions as panel chairs from ∼0.1 of the total number of chairs to ∼0.45 by AO-21.

4.6   XMM-Newton User Group

As well as the OTAC, there is another body involved in optimising the scientific output of the mission. This is the XUG which advises ESA, through the Project Scientist, on all matters relating to the optimisation of the scientific output of the mission. It also acts as a forum to discuss input from the community of users, and when appropriate advise or recommend action to ESA regarding XMM-Newton operations. There were 15 members (including the chair) when the XUG was established in 2002 which decreased to ten members in 2011. The chair of the OTAC is also a member of the XUG to ensure good coordination between the two bodies. In addition, the XMM-Newton Project Scientist, Mission Manager, Science Support Manager and instrument principal investigators attend parts of the meetings. The names of all the members are made public. Genders were assigned in the same way as for the OTAC. Figure 4.4 shows the number of female XUG members divided by the total between 2002 and 2021. An notable increase in the fraction of female members from ∼5% to ∼40% of the total is evident.

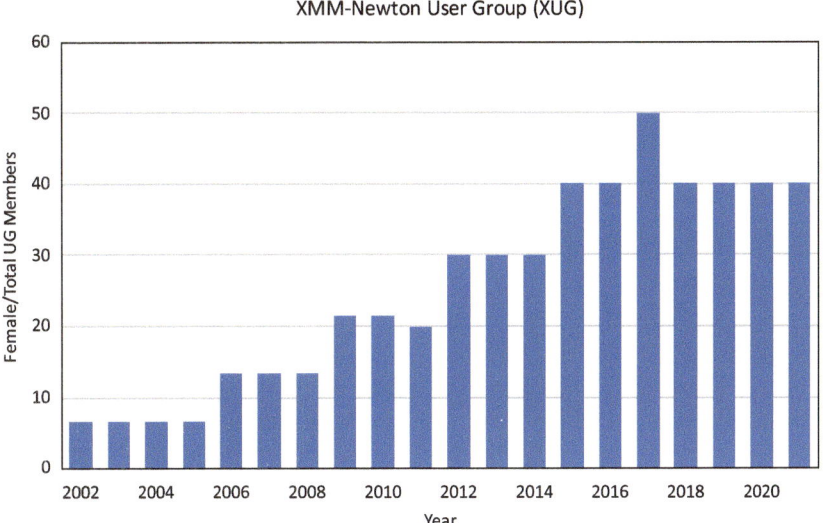

Fig. 4.4 The percentage of female XMM-Newton XUG members compared to the total between 2002 and 2021. An increase in the percentage of female members from ~5% to ~40% of the total is evident

4.7   Proposers

Between AO-2 and AO-21 there were 10,579 proposals submitted in response to XMM-Newton calls for observing opportunities. This provides a rich data set to quantify many aspects of the proposal process. When submitting a proposal, XMM-Newton proposers are required to indicate the country where their institute or university is located using a drop-down menu. This information allows the nationalities of the proposers' institutes to be examined. Proposers are not required to submit information on gender, age or type of position held and these have to be derived by other means if required.

4.7.1   Proposers' Institute Countries

Table 4.2 shows the number of proposals submitted and accepted for a range of countries including all 22 ESA Member States and other countries from which significant numbers of proposals originated. Accepted proposals are ones which were awarded any observing time. Only 31 proposals were received from countries that are not listed individually in Table 4.2.

Scientists located at institutes within the United States submitted the most proposals – 40.7% of the total. This is followed by Germany (11.9%), Italy (11.5%), the United Kingdom (8.9%), Spain (5.7%) and France (3.7%). Figure 4.5 shows

Table 4.2 The number of proposals submitted and accepted from PIs located in a range of countries. In this case "accepted" means awarded any observing time

Country	No. proposals submitted	Percentage of total	No. accepted	Acceptance percentage
AUSTRIA	26	0.25	6	23.1
BELGIUM	150	1.42	57	38.0
CZECH REPUBLIC	15	0.14	5	33.3
DENMARK	9	0.09	5	55.6
ESTONIA	4	0.04	0	0.0
FINLAND	49	0.46	16	32.7
FRANCE	394	3.72	189	48.0
GERMANY	1258	11.89	535	42.5
GREECE	40	0.38	12	30.0
HUNGARY	6	0.06	0	0.0
IRELAND	32	0.30	6	18.8
ITALY	1215	11.49	538	44.3
LUXEMBOURG	0	0.00	0	0.0
NETHERLANDS	361	3.41	185	51.2
NORWAY	1	0.01	1	100.0
POLAND	34	0.32	12	35.3
PORTUGAL	4	0.04	1	25.0
ROMANIA	0	0.00	0	0.0
SPAIN	598	5.65	238	39.8
SWEDEN	17	0.16	6	35.3
SWITZERLAND	115	1.09	50	43.5
UNITED KINGDOM	941	8.89	406	43.1
ARGENTINA	14	0.13	5	35.7
AUSTRALIA	43	0.41	17	39.5
BRAZIL	18	0.17	10	55.6
BULGARIA	13	0.12	6	46.2
CANADA	161	1.52	65	40.4
CHILE	35	0.33	12	34.3
CHINA	116	1.10	31	26.7
INDIA	83	0.78	20	24.1
ISRAEL	28	0.26	12	42.9
JAPAN	304	2.87	90	29.6
KOREA	20	0.19	5	25.0
MEXICO	35	0.33	9	25.7
RUSSIA	28	0.26	9	32.1
TAIWAN	46	0.43	10	21.7
TURKEY	29	0.27	10	34.5
UNITED STATES	4306	40.70	1699	39.5
OTHER	31	0.29	9	29.0

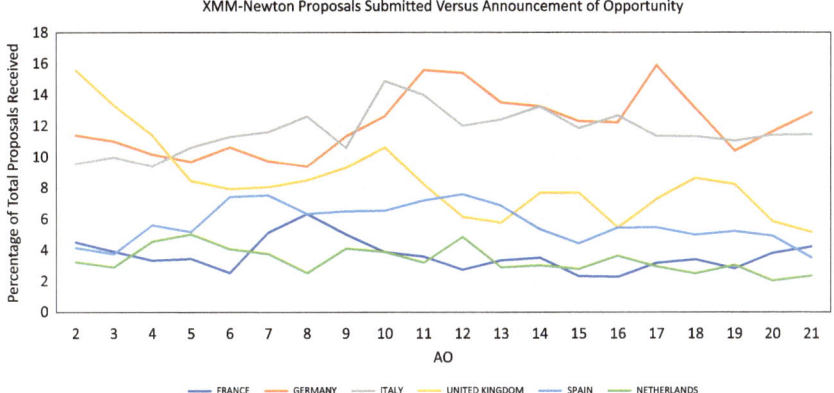

Fig. 4.5 Percentages of proposals submitted from the six ESA Member States with the largest numbers of proposals between AO-2 to AO-21. A decline in the percentage of proposals from the United Kingdom from ∼15% for the earliest AOs to ∼6% for the latest is evident

the variations in percentage of proposal numbers versus AO for six ESA Member States (Germany, France, Italy, the Netherlands, Spain and the United Kingdom). These show relatively stable proposal submission fractions with AO number with the possible exception of the United Kingdom which shows a decline from ∼15% for the earliest AOs to ∼6% for the latest. There is evidence for a small increase in the submission fraction with time of proposers located in Germany.

Figure 4.6 shows the variations in percentage of proposal numbers versus AO for four Non-Member States (Canada, China, India and Japan). China shows an increase of the proposal submission fractions with AO number from 0% to ∼3%. A similar, but weaker, trend is seen for India where the submission fractions increases from 0% to ∼1.5%. The submission fractions with AO number of Japan shows a steep decline from ∼7% for AO-2 to ∼2% in AO-4 and stays stable at this level for all later AOs. The decline may reflect the successful launch of the Japanese Aerospace Exploration Agency (JAXA)/NASA X-ray mission Suzaku and its scientific exploitation in the subsequent years.

In order to be able to draw reliable conclusions, we examined the number of accepted proposals for the 13 countries whose scientists have submitted >80 proposals (Table 4.2). The sum of all the submitted proposals from these countries is 95% of the total number of submitted proposals. The countries with the highest accepted fractions are the Netherlands where 51.2% of 361 proposals were accepted, followed by France with 48.0% of 394 proposals, Italy with 44.3% of 1215 proposals, Switzerland with 43.5% of 115 proposals, the United Kingdom with 43.1% of 941 proposals and Germany with 42.5% of 1258 proposals. This may reflect the well-established high-energy communities in these countries.

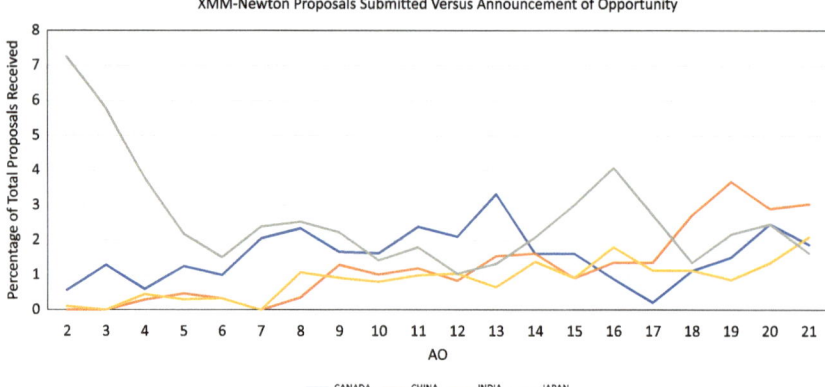

Fig. 4.6 Percentages of proposals submitted from the Non-Member States with the largest numbers of proposals (excluding the USA) between AO-2 to AO-21. China and India show increases with AO number from 0% to ~3% and from 0% to ~1.5%, respectively. The large decrease in the percentage of proposals from Japan may be related to the launch of the JAXA/NASA Suzaku X-ray mission and its subsequent exploitation

Table 4.3 The number of proposals submitted and accepted from male and female PIs located in countries with >80 proposals in total. The differences are with respect to the acceptance percentages given in Table 4.2 which are for all proposals

Country	Proposals submitted		% Female	Proposals accepted		% Difference	
	Male PI	Fem. PI	/Total	Male PI	Fem. PI	Male PI	Fem. PI
BELGIUM	86	64	42.7	35	22	2.7	−3.6
FRANCE	258	136	34.5	141	48	6.7	−12.7
GERMANY	945	313	24.9	413	122	1.2	−3.6
ITALY	843	372	30.6	389	149	1.9	−4.2
NETHERLANDS	267	94	26.0	130	55	−2.6	7.3
SPAIN	431	167	27.9	175	63	0.8	−2.1
SWITZERLAND	96	19	16.5	45	5	3.4	−17.2
UNITED KINGDOM	752	189	20.1	322	84	−0.3	1.3
CANADA	80	81	50.3	33	32	0.9	−0.9
CHINA	103	13	11.2	31	0	3.4	−26.7
INDIA	60	23	27.7	17	3	4.2	−11.1
JAPAN	234	70	23.0	66	24	−1.4	4.7
UNITED STATES	3400	906	21.0	1374	325	1.0	−3.6

4.7.2 Proposers' Gender

We next examined the number of proposals with male and female PIs from these 13 countries (Table 4.3). Since proposers are not asked to specify their gender (nor age or type of position held etc.) gender information for each proposer was obtained through examining publicly-accessible web-based data in a similar way as for the

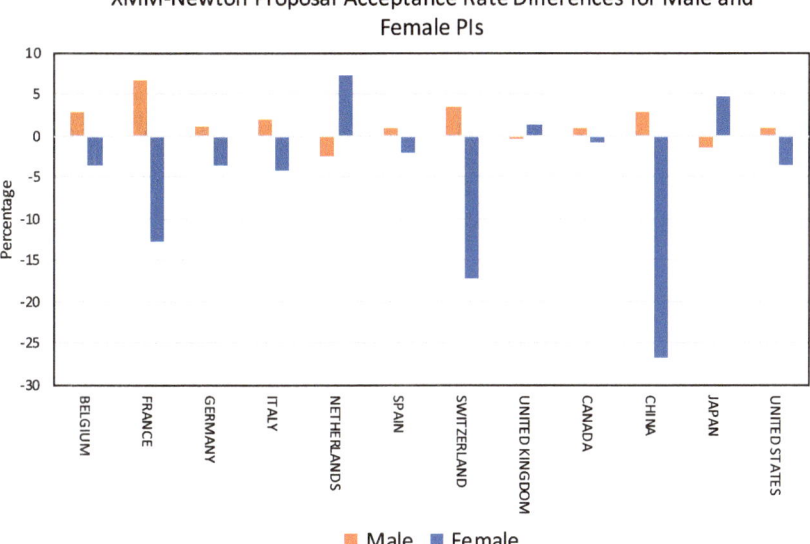

Fig. 4.7 The acceptance fraction differences of XMM-Newton proposals with male and female PIs compared to the average for all proposers from that country. The countries are those with >100 submitted XMM-Newton proposals

HST study by I.N. Reid [68]. We appreciate that gender identity is more complex than a binary issue, however, no attempt can be made to assign genders other than male or female as this information is not readily available.

Among the countries with more than 80 proposals submitted, the country with the highest fraction of female PIs is Canada (50.3%) followed by Belgium (42.7%). The countries with the lowest fraction of female PIs are China (11.2%) and Switzerland (16.5%). Figure 4.7 shows the acceptance fraction differences for male and female PIs compared to the average for that country for each of the 12 countries with >100 proposals. The two countries with the lowest fraction of submitted proposals from female PIs (China and Switzerland) also have the lowest fractions of proposals with female PIs accepted of -26.7% and -17.2%, respectively, compared to the average. Only three countries have better female PI acceptance rates than for male PIs – the Netherlands, Japan and the United Kingdom with 7.3, 4.7, and 1.3%, respectively compared to the averages.

Figure 4.8 shows the fraction of XMM-Newton proposals with female PIs compared to the total. This increased relatively steadily from ∼20% to ∼30% between AO-2 and AO-21 (2001 to 2021).

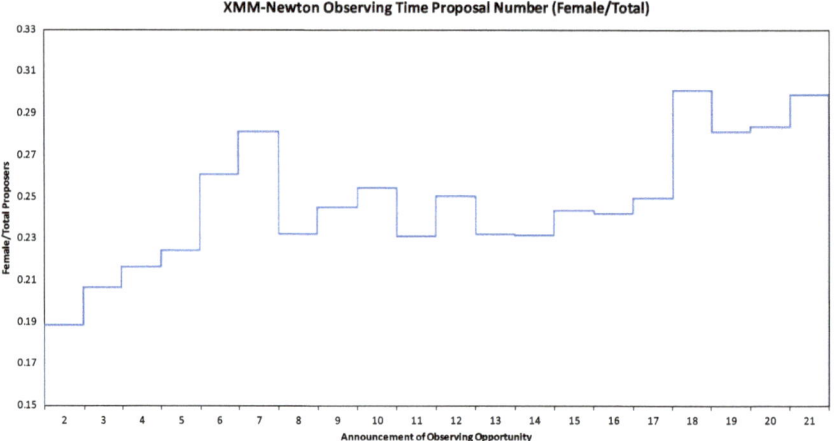

Fig. 4.8 The fraction of submitted proposals with a female PI compared to the total between AO-2 to AO-21 (2001 to 2021). Results are for all countries where PIs submitted proposals

4.7.3 Proposers' Academic Age

In order to determine the "academic age" of proposers we used the difference between the year that their Doctor of Philosophy Degree (PhD) was awarded and the year that an AO to which they submitted a proposal was issued. Thus a proposer who submits multiple proposals to the same AO will be counted multiple times in the same "academic year", whilst one who submits multiple proposals to different AOs will be counted in different "academic years".

The year a proposer obtained their PhD (or equivalent) was determined for 94.8% of the proposals by searching the internet, particularly sites such as the Astrophysics Data Service (ADS), LinkedIn, the Astronomy Genealogy Project (astrogen.aas.org), ORCID.org, IEEE Xplore (https://ieeexplore.ieee.org/Xplore/home.jsp), and, for French theses https://www.theses.fr. Some proposers were contacted directly and provided their PhD dates by email. The proposers for which the PhD could not be found are often retired, deceased or have left astronomy. For PhD students who had not yet completed their degrees, the expected year of submission was used. For the small number of proposers who did not have a PhD and were not enrolled in a PhD programme, their dates were assumed to be arbitrarily far in the future. For some late career scientists in Italian institutes who do not have a PhD, their "academic age" was taken to be three years after they obtained their Laurea. It should be noted that using the year of PhD to indicate the number of years experience neglects time spent outside of astronomy. We note that the "youngest" proposer was 13 years before obtaining a PhD and the oldest 52 years after! The mean "academic age" of PIs is 10.9 years. For female PIs it is 9.8 years post-PhD, compared to 11.2 years for male PIs (Figs. 4.9, 4.10, 4.11, and 4.12).

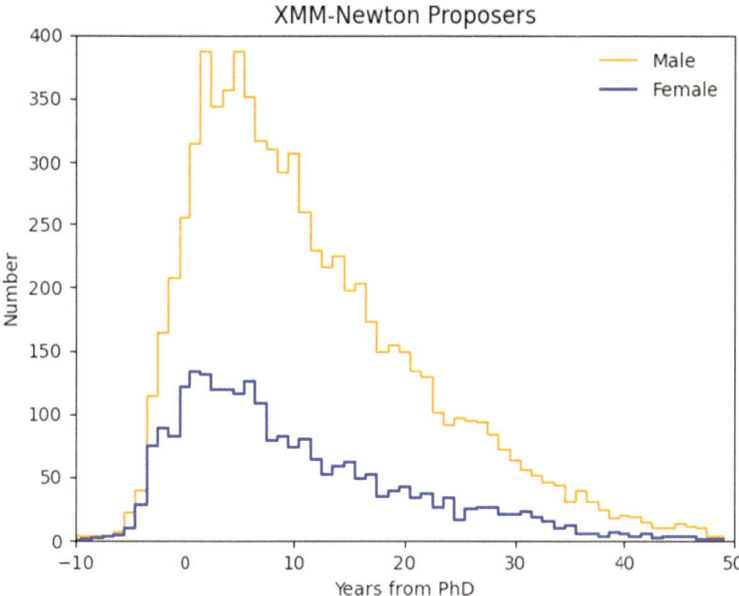

Fig. 4.9 The "academic age" distribution (years since PhD) for male and female XMM-Newton PIs between −10 and 50 years. The mean "academic age" PIs is 10.9 years. For female PIs this is 9.8 years post-PhD, compared to 11.2 years for male PIs

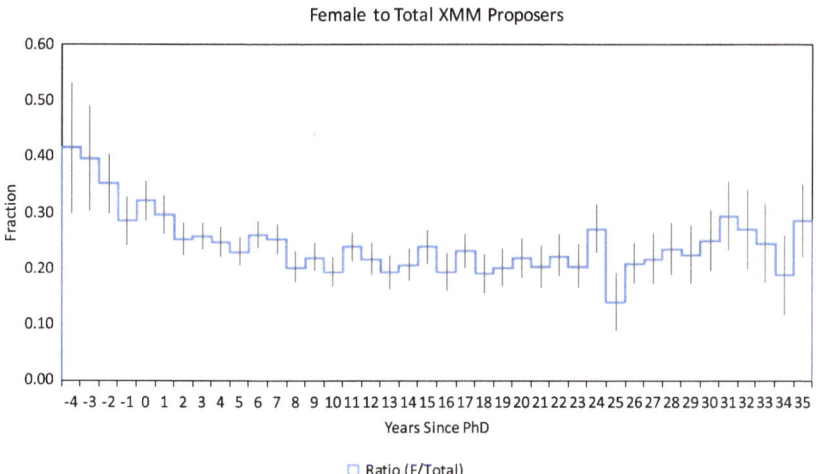

Fig. 4.10 The ratio of the "academic age" (years since PhD) of female compared to all XMM-Newton PIs showing the increase in early career (less than around 5 years from obtaining a PhD) female PIs compared to the total. The indicative error bars show 1σ standard deviations assuming that the number of proposers in each age bin follows a Poisson distribution

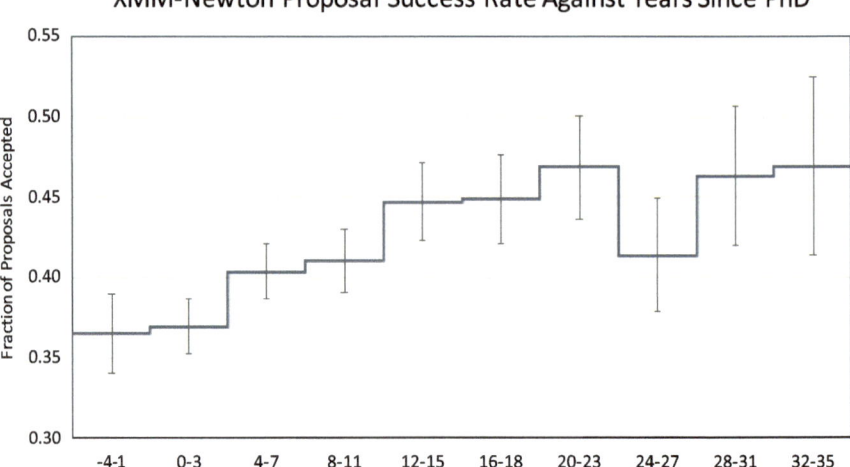

Fig. 4.11 The acceptance fraction of XMM-Newton PIs with year of their PhD between −4 and 35 years after PhD. The indicative error bars indicate 1σ standard deviations assuming that the number of proposers in each age bin follows a Poisson distribution

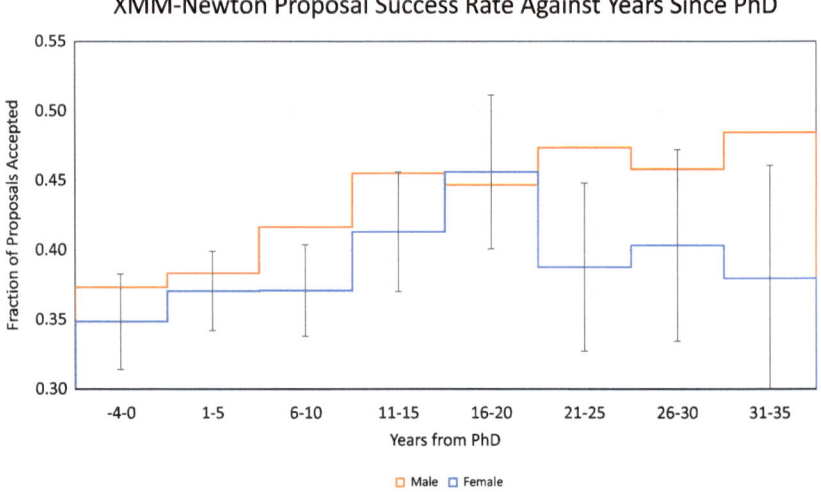

Fig. 4.12 The acceptance fractions of male and female XMM-Newton PIs with year of their PhD between −4 and 35 years after PhD. The success rate of male PIs shows a gradual increase with "academic age" from ∼0.38 to nearly 0.50 over the age range considered. For female PIs the success rate may increase in a similar manner to their male colleagues until ∼20 years after PhD, after which the increase appears to stop and may even reverse. The indicative error bars indicate 1σ standard deviations assuming that the number of proposers in each age bin follows a Poisson distribution. Error bars are only shown for female PIs as these are much larger than for male PIs due to the smaller number of female PIs

4.8 Proposal Selection

We evaluated the outcomes of the XMM-Newton AO-2 to AO-21 selection processes for both genders using two different methods:

1. Proposals that were awarded priority A, B or C observing time.
2. Proposals that were awarded high-priority (A or B) observing time.

Observations from high-priority proposals (A or B) are guaranteed to be performed, so having an approved high-priority proposal is more likely to result in scientific papers and so be beneficial for the career of a researcher. In contrast only around 40% of priority C proposals are actually observed. Since many XMM-Newton proposals are allocated less observing time than requested, almost always because not all the requested targets are approved, we decided to examine both the numbers of successful proposals and how much observing time was awarded. We note that in the later AOs there was significant time allocated to Large and Multi-Year Heritage Programmes that required substantial investments in observing time. If the allocation of observing time is handled differently for males and female PIs, then this will show as a difference in the relative amounts of time approved, compared to the number of proposals accepted. Thus, for both outcomes, the numbers and amounts of awarded observing times were determined for the total, male PI and female PI populations (Tables 4.4 and 4.5).

4.8.1 Proposal Selection: Priority A, B or C

We first examined the number of proposals that were awarded *any* observing time. A total of 10,579 proposals were submitted to XMM-Newton AO-2 to AO-21 of which 4287 were awarded observing time. This corresponds to an overall success rate of 40.5%. A total of 7997 proposals with male PIs and 2582 proposals with female PIs were submitted, of which 3316 and 971 proposals were awarded *any* observing time in priority A, B, or C. This gives success rates of 41.5 and 37.7% for male and female PI proposals, respectively. This is a difference in favour of males of 10.1%.

Proposal submissions and proposal acceptance are unlikely to be independent or random processes and for Poisson statistics to apply events need to be independent of each other. However, to illustrate the uncertainties that would apply if such statistics are applicable, we used square root uncertainties to find success rates of (41.5 ± 1.0) and $(37.7 \pm 1.8)\%$ for male and female PI proposals, respectively. This is a difference in favour of males of $(10.1 \pm 5.9)\%$ which is formally significant at 1.7σ. We emphasise that these uncertainties are only given to illustrate the outcomes if Poisson statistics were to apply to the selection process. We note that actual uncertainties on the proposal numbers for each AO are zero. We further note that there are systematic uncertainties associated with this process due to

Table 4.4 XMM-Newton AO proposal numbers showing for proposals with male and female PIs and the number of proposals accepted with high-priority (Priority A and B) and for all accepted (Priority A, B and C)

	Submitted			Accepted			
	No. of	Male led	Female led	Priority A+B		Priority A+B+C	
AO	proposals	proposals	proposals	Male	Female	Male	Female
2	869	705	164	181	40	302	61
3	692	549	143	196	63	291	80
4	660	517	143	146	32	182	46
5	641	497	144	135	33	210	54
6	594	439	155	127	18	193	46
7	586	421	165	97	32	146	52
8	555	426	129	88	23	135	43
9	539	407	132	91	28	151	46
10	491	366	125	90	26	168	49
11	501	385	116	74	23	146	49
12	475	356	119	62	24	128	50
13	452	346	106	87	21	162	43
14	431	331	100	78	19	155	33
15	431	326	105	71	22	151	45
16	442	335	107	86	17	148	35
17	441	331	110	67	20	126	43
18	442	309	133	64	25	128	47
19	462	331	131	63	26	132	50
20	447	320	127	71	22	142	44
21	428	300	128	55	24	129	55
Total	**10,579**	**7997**	**2582**	**1929**	**538**	**3316**	**971**

misappropriated genders and incorrect PhD "academic ages", but these are likely to be too small to significantly affect our calculations.

The success rates are shown for male, female and all proposers for each AO in Fig. 4.13. In contrast to the results from HST Cycles 11 to 20 reported in [68] where proposals with male PIs had a consistently higher success rate than those with female PIs, XMM-Newton proposals with male PIs had higher success rates for 14 of the AOs and female PIs had higher success rates for 5 AOs. For AO-21, the success rates are almost identical. Interestingly, there is no obvious evolution in the male and female PI proposal acceptance rates with AO number over an interval of more than 20 years.

Figure 4.14 shows the variation in acceptance rates more clearly. For each AO, it shows the difference between the expected number of accepted proposals, calculated using the overall acceptance rate, and the number actually accepted. It shows that the largest discrepancies occurred during AO-6 and AO-14 where proposals with a female PIs had lower success rates of 16.4 and 10.6%, respectively, compared to proposals with a male PI. The most successful AO for female PI proposals was AO-

Table 4.5 XMM-Newton requested and accepted observing time for proposals with female and male PIs between AO-2 and AO-21 for high priority (A or B) and Priority A, B, and C. Time is in units of Msec

| | Req. Time | | Priority A+B | | | | Priority A+B+C | | | |
| | | | Acc. Time | | Success Rate | | Acc. Time | | Success Rate | |
AO	Male	Female	Male	Female	Male	Female	Male	Female	Male	Female
2	96.5	21.7	12.8	3.0	0.132	0.136	21.5	4.6	0.223	0.212
3	81.9	16.9	15.5	3.9	0.190	0.231	26.8	6.0	0.327	0.353
4	80.8	21.1	11.8	3.1	0.146	0.145	15.2	4.1	0.187	0.196
5	87.3	19.5	9.6	3.1	0.110	0.159	16.3	4.8	0.187	0.245
6	77.0	23.1	11.7	1.2	0.152	0.051	17.3	3.3	0.225	0.143
7	87.8	26.0	9.6	3.5	0.109	0.136	14.2	6.8	0.161	0.263
8	96.6	21.5	11.4	1.7	0.118	0.077	17.6	4.3	0.182	0.199
9	87.7	26.5	10.2	2.6	0.117	0.100	16.1	4.8	0.184	0.180
10	66.6	22.9	9.1	5.3	0.137	0.230	17.3	7.7	0.259	0.338
11	66.4	21.1	9.4	2.5	0.142	0.118	15.6	6.0	0.235	0.286
12	61.5	24.4	9.5	2.4	0.154	0.098	15.8	5.0	0.257	0.204
13	58.9	18.8	10.8	3.7	0.183	0.197	18.2	6.1	0.309	0.325
14	60.8	24.1	11.7	2.2	0.193	0.103	21.3	3.7	0.350	0.173
15	56.5	24.1	10.1	3.1	0.179	0.129	19.0	5.4	0.337	0.223
16	65.9	25.2	12.7	2.1	0.193	0.085	21.1	4.0	0.320	0.158
17	96.3	38.5	11.6	5.3	0.120	0.137	20.0	7.7	0.208	0.199
18	63.7	29.3	8.0	3.9	0.125	0.132	16.2	9.4	0.254	0.320
19	66.8	25.1	7.9	2.9	0.118	0.117	17.6	6.7	0.264	0.268
20	76.1	30.6	15.4	2.3	0.202	0.074	24.3	6.1	0.319	0.198
21	56.7	24.3	8.5	3.1	0.151	0.127	17.0	7.5	0.300	0.307
Total	**1491.8**	**481.6**	**217.3**	**60.8**			**368.4**	**113.8**		

12 where female-led proposals were 5.4% more likely to be accepted than ones with a male PI (Fig. 4.15).

We examined the OTAC membership genders to see if these are correlated with proposal acceptance rates. Figures 4.16 and 4.17 show the acceptance fraction of priority A and B proposals with female PIs against the fractions of female compared to the total number of OTAC members and the fraction of female OTAC panel chairs, respectively. It can be seen that there is no obvious correlation between the success rates of female-led proposals and the fraction of female OTAC members or panel chairs.

The female PI success rate during AO-3 of 0.44 was unusually high, but based on only 63 proposals accepted with Priority A or B. There were no female panel chairs for AO-5 and AO-6. AO-5 has an average female PI Priority A and B success rate of 0.23, whereas AO-6 has the lowest female PI success rate of any of the AOs of 0.12 (Fig. 4.3). This supports the view that having females in leadership positions in the OTAC may improve female PI proposal success rates.

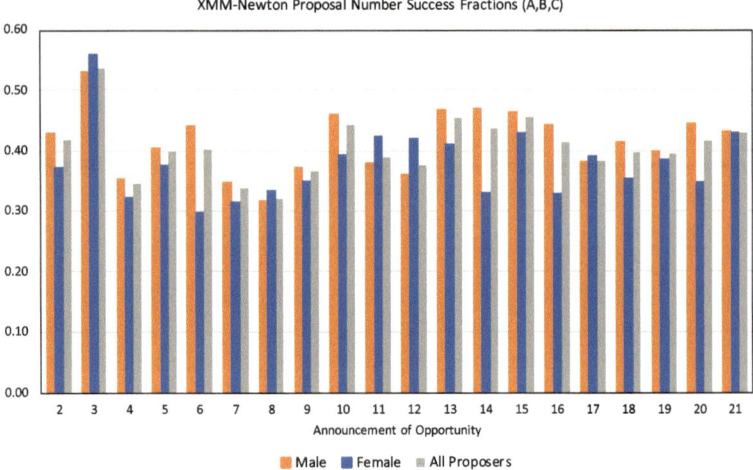

Fig. 4.13 The fraction of accepted (Priority A, B or C) proposals compared to the number submitted. The histograms show the acceptance fractions for male PIs (orange), female PIs (blue) and all PIs (grey)

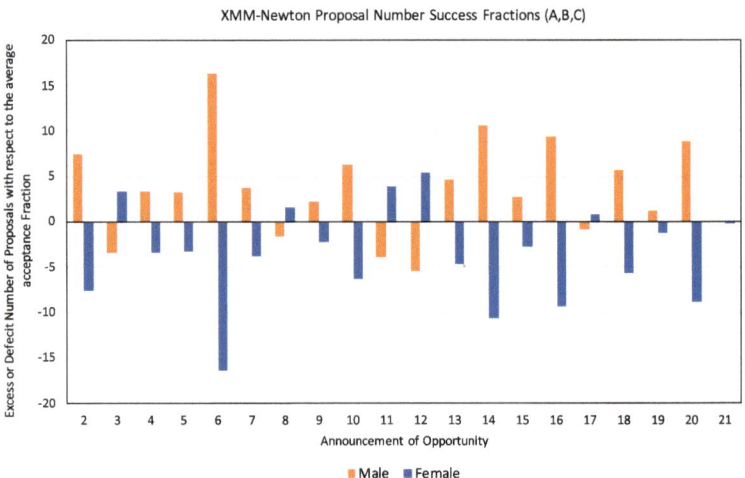

Fig. 4.14 The relative success rates of XMM-Newton priority A, B and C proposals. The histograms show the difference between the actual number of successful proposals and the expected number based on the overall acceptance rate

Figure 4.15 shows the same excesses normalised by dividing by the square root of the expected number of accepted proposals. As discussed earlier, Poisson statistics do not necessarily apply to the peer-review process. Indeed, if the individual results are examined, the individual results could arise from sampling statistics of an underlying distribution with equal male and female success rates. However, such an

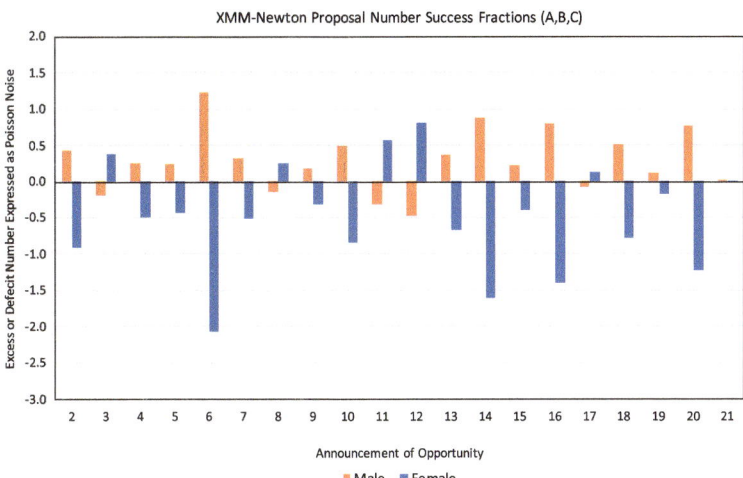

Fig. 4.15 The relative success rates of XMM-Newton priority A, B and C proposals. The histograms show the difference between the actual number of successful proposals and the expected number based on the overall acceptance rate normalised by dividing by the square root of the number of expected proposals

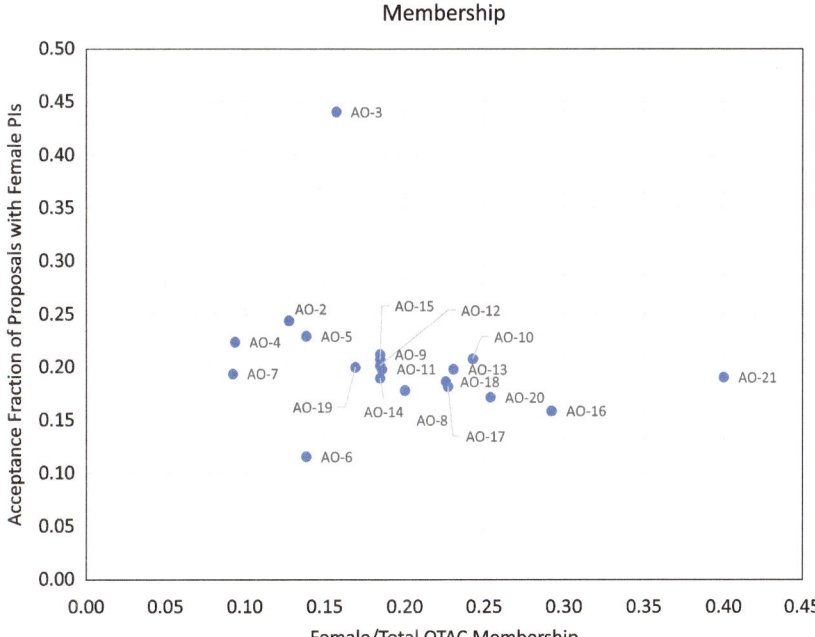

Fig. 4.16 The acceptance fraction for female PIs for each AO (labelled) against the total fraction of female OTAC members. There is no obvious correlation between the two

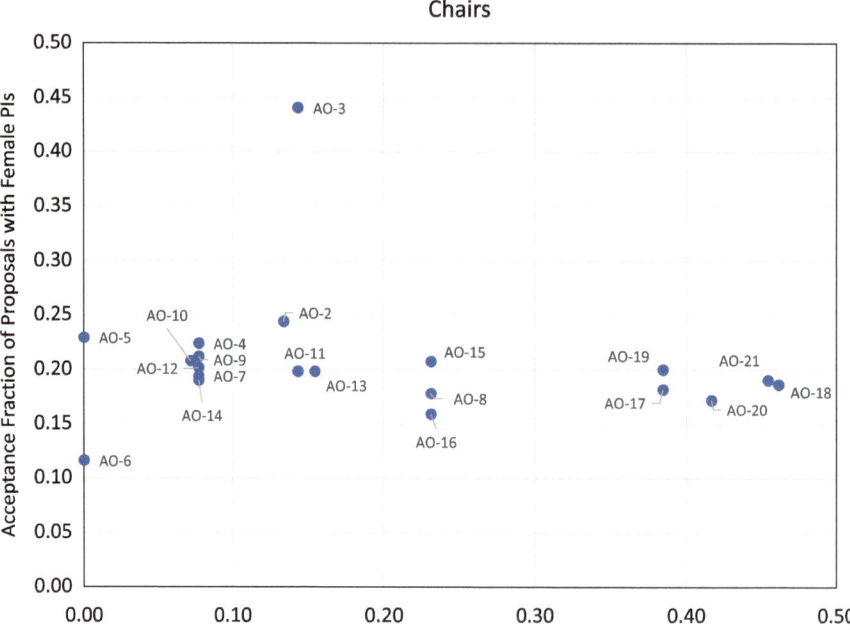

Female Proposal Priority A,B Acceptance Fraction versus OTAC Panel Chairs

Fig. 4.17 The acceptance fraction for female PIs for each AO (labelled) against the fraction of female OTAC panel chairs. There is no obvious correlation between the two

argument is more difficult to sustain when viewed in the context of the distribution of results from all 20 AOs.

We next examined the observing times awarded in priority A, B and C. The average success rate for time is 24.7% for male PIs and 23.6% for female PIs of that requested (Table 4.7). This is a difference of 4.5% in favour of male PIs. This smaller difference when the number of proposals are compared of 10.1% and may suggest that female-led proposers are awarded relatively more observing time than their male counterparts.

These results indicate that proposals with male PIs are more likely to be approved than those with female PIs. This prompted us to examine whether differences in the population of XMM-Newton male and female observing time proposers could account for this difference. This could happen for example if the underlying male population is more experienced and if the proposal success rate increases with experience as would seem likely.

Figure 4.9 shows the numbers of male and female XMM-Newton PIs who obtained their PhDs or equivalent in five year bins between 4 years before the PhD date to 35 years after. A two-sample Anderson-Darling Test (e.g., [69]) shows that the hypothesis that both samples come from the same underlying population can be rejected at >99% confidence. The differences between the two samples can be more clearly seen in Fig. 4.10 which shows the ratio of the male and female PI distributions. There is an increase in the number of early career (less than around 5 years from obtaining a PhD) female PIs compared to males; beyond ~10 years after PhD, the ratio of females to all proposers is ~20%, compared to ~40% for the early career years.

We next investigated the acceptance rates for proposals from all proposers (male and female PIs) against PhD year. This is shown plotted for years −4 to 35 years after PhD in Fig. 4.11. A linear fit to the data gives an intercept of $(37.3 \pm 0.4)\%$ and a gradient of $(0.35 \pm 0.15)\%$ year^{-1} assuming that the uncertainties are the square root of the number of people in each bin. The value of R^2 is 0.35. This is consistent with an increase from a ~35% acceptance rate for PhD students to ~45% for senior researchers. Such an increase could result from a combination of factors including:

1. An increase in the success rate of researchers as they gain experience and have expanded networks of collaborators.
2. Less successful proposers who leave astronomy.
3. Less successful proposers who remain in astronomy, but do not propose in subsequent XMM-Newton AOs.
4. A bias in the selection process towards late career scientists.

We note that the measurements are also consistent with an acceptance rate that remains approximately constant beyond ~12 years post-PhD and decreases more strongly for lower "academic ages".

The positive gradient of this relation indicates that the female proposer population is less likely to have proposals accepted simply due to having less experience. To calculate the size of this effect we took the male and female PI distributions shown in Fig. 4.9 and multiplied the number of PIs each year by the linear value derived from the fit to the overall success rate. This gave a predicted difference in the acceptance ratio of 1.0% between male and female PIs. This indicates that the underlying "academic age" differences between the two populations is probably not responsible for the majority of the differences in acceptance rate.

The acceptance fractions of male and female XMM-Newton PIs with year of their PhD are shown in Fig. 4.12. Proposals from male PIs have higher success rates in seven out of the eight bins. The success rate of proposals from male PIs is consistent with a gradual increase with "academic age" throughout the range examined (−4 to 35 years from PhD) from around 0.38 to nearly 0.50. For proposals with female PIs the success rate is consistent with increasing in a similar manner to their male counterparts until ~20 years post PhD, after which the gradual increase appears to stop and may reverse. We note that a similar decrease for senior female PIs is evident in HST Cycles 11 to 21 data [68]. We cannot exclude that proposals from female PIs simply have an average acceptance rate of ~0.39, independent of "academic age".

4.8.2   Proposal Selection: Priority A or B

We then investigated the outcomes of the proposals that were accepted with Priority A or B. Observations from these proposals are guaranteed to be performed, so having an approved high-priority proposal is likely to result in scientific papers and so be beneficial for the career of a researcher. This allows the investigation of any gender dependence of acceptance with assigned proposal priority. We note that an average of 42% of Priority C proposal targets were observed. Of the total of 10,579 proposals, 2467 were awarded observing time in priority A or B corresponding to an acceptance rate of 23.3%. Of the total of 7997 male-led and 2582 female-led proposals that were submitted, 1929 and 538 proposals were awarded observing time in priority A or B. This gives success rates of 24.1 and 20.9% for proposals with male and female PIs, respectively. This is a difference in favour of males of 15.6%. If we assume square root errors on the numbers of submitted and accepted proposals this implies success rates of (24.1 ± 0.6) and $(20.9 \pm 1.0)\%$ for proposals with male and female PIs, respectively and a difference in favour of males of $(15.6 \pm 6.2)\%$ which is significant at 2.5σ. Square root uncertainties are unlikely to apply to the selection process and are only used to illustrate outcomes should this be the case.

This difference is higher than when Priority C time is included. This implies that the OTAC approves relatively more Priority C proposals with female PIs compared to males PIs. The success rates are shown for male, female and all proposers in Fig. 4.18. Again, there is no obvious evolution in the male and female PI proposal acceptance rates with AO number, which cover an interval of more than 20 years. Figure 4.19 shows the variation in high-priority acceptance rates more clearly. For each AO, it shows the difference between the expected number of accepted proposals, calculated using the overall acceptance rate, and the number actually accepted. It shows that the largest discrepancies occurred during AO-6 and AO-16 where proposals with a female PI had lower success rates of 19.9% and 7.9%, respectively, compared to proposals with a male PI. The most successful AO for proposals with a female PI was AO-3 where such proposals were 9.5% more likely to be accepted than ones with a male PI. Figure 4.20 shows the same excesses normalised by dividing by the square root of the expected number of accepted proposals.

The average success rates for high priority (A or B) time is 14.6% of the requested times for male PIs and 12.6% for female PIs. This is a difference of 15.4% in favour of male proposers, which is very similar to the difference in the number of accepted high-priority proposals. This indicates that the assignment of high-priority observing time is handled similarly for proposals with male and female PIs (Fig. 4.5).

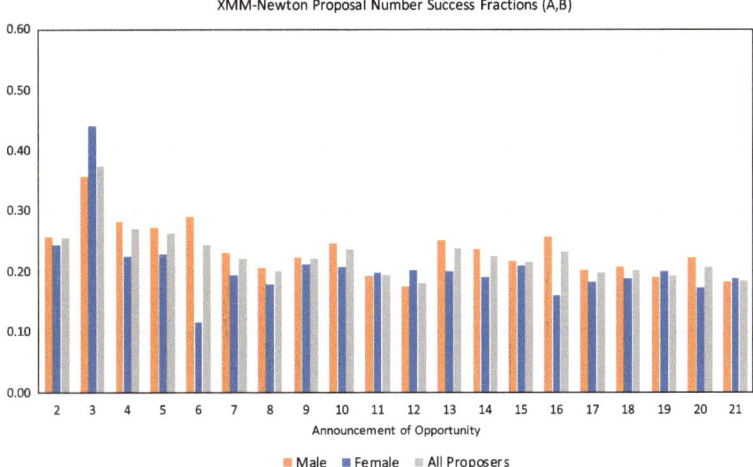

Fig. 4.18 The fraction of high-priority (A or B) proposals compared to the number submitted. The histograms show the success rates for male PI (orange), female PI (blue) and all PIs (grey)

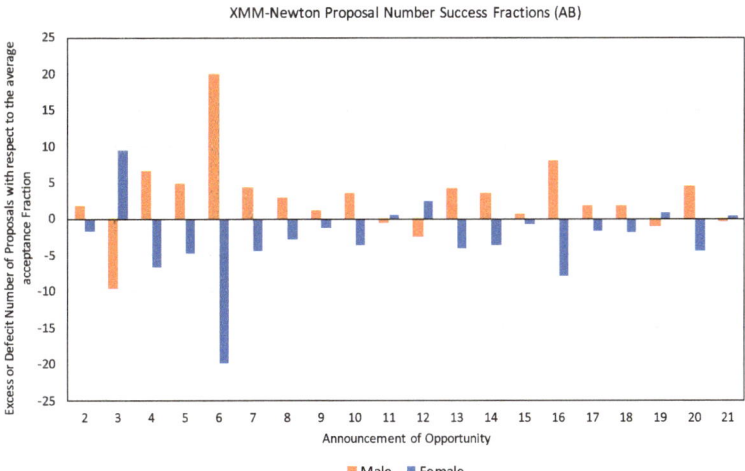

Fig. 4.19 The relative success rates of XMM-Newton high-priority proposals. The histograms show the difference between the actual number of successful proposals and the expected number based on the average acceptance rates

4.8.3 Proposal Selection: Outcomes

We have examined the outcomes of 10,579 proposals submitted to ESA in response to the AO-2 to AO-21 calls for XMM-Newton observing time which were released between 2001 and 2021. The requested over-subscription in observing time was high ($\gtrsim$5) during all the AOs. During this time there was an increase in the fraction

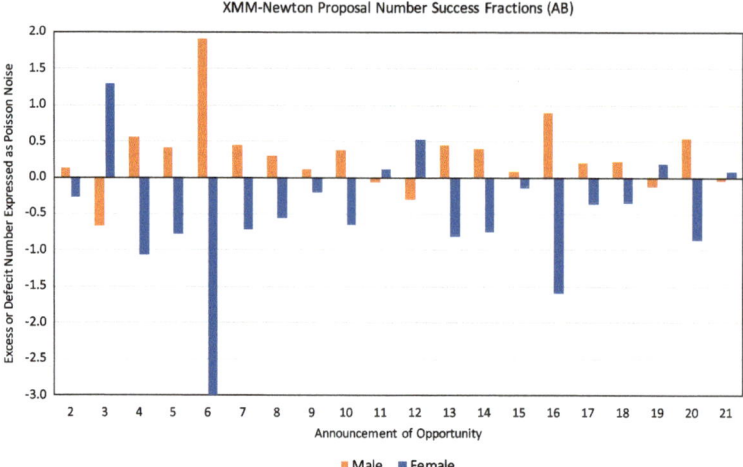

Fig. 4.20 The relative success rates of high-priority XMM-Newton proposals. The histograms show the difference between the actual number of successful proposals and the expected number based on the overall acceptance rate normalised by dividing by the square root of the number of expected proposals

of female OTAC panel members from ∼15% to ∼25% of the total. The increase in female panel chairs is more marked from ∼10% to ∼45% of the total chairs by AO-21. Similarly, the fraction of female XUG members increased from ∼5% to ∼40% of the total. There is no obvious correlation between female PI success rates and the fraction of female OTAC members or panel chairs. We note that the lowest female success rate (0.143 during AO-6 for Priority A, B and C) occurred in one of only two AOs where the OTAC did not have any female panel chairs.

Scientists located at institutes within the USA submitted the most proposals – 40% of the total followed by those located in Germany, Italy and the UK. The Netherlands has the highest fraction of accepted proposals amongst the 13 countries submitting >80 proposals. Amongst these countries, Canada and Belgium have the highest fractions of proposals from female PIs; 50.3 and 42.7%, respectively. China and Switzerland have the lowest fractions; 11.2 and 16.5%, respectively. Three countries have better female PI proposal acceptance rates than for male PIs – the Netherlands, Japan and the United Kingdom with 7.3, 4.7, and 1.3%, respectively compared to their averages.

The fraction of XMM-Newton proposals from female PIs increased from ∼20% of the total during the early AOs to ∼30% of the total in the latest AOs. 41.5% of proposals with male PIs and 37.7% with female PIs were awarded any observing time on XMM-Newton. This is a difference in favour of males of 10.0%. There is no marked evolution in gender acceptance rates with AO number. The difference in success rates with gender is more marked when only proposals that were ranked to be high-priority (Priority A and B) are considered. These proposals are guaranteed to have their approved targets observed. The acceptance rate is 24.1% for proposals with male PIs and 20.9% for those with female PIs. This is a difference of 15.6%. This suggests that the OTAC ranks relatively more proposals with female PIs with a lower Priority C, than proposals with male PIs.

In order to investigate whether the OTAC awarded observing time differently for the two genders, the amounts of observing time awarded were also examined. For the high-priority proposals (priority A and B) 14.6% of the requested time was awarded for proposals with a male PI and 12.6% for proposals with a female PI. This is a difference of 15.4% in favour of male PIs. Similarly, for all the proposals, 24.7% of the requested time was awarded for proposals with a male PI and 23.6% for female PIs. The difference is 4.5% in favour of male PIs, smaller than the 10% difference when proposal numbers are considered. Again this implies that female PIs are relatively more likely to be awarded Priority C observing time. These results are summarised in Table 4.7. We interpret the different male/female PI proposal acceptance rates to imply that male PIs are between 5–15% more likely to benefit in the XMM-Newton proposal assessment process that their female counterparts.

Using the year of obtaining a PhD, or equivalent degree gives the mean "academic age" of PIs as 10.9 years post-PhD. For female PIs this is 9.8 years post-PhD compared to 11.2 years for male PIs. A linear fit to the proposal (Priority A, B and C) number acceptance rate against PhD year shows an increase in acceptance percentage from $\sim$35% prior to PhD to $\sim$45% for more "senior" astronomers. This increase in proposal acceptance probability with "academic age" favours a senior male population. However, this effect only accounts for $\sim$1% of the 5–15% difference in acceptance rates between male and female PIs.

4.8.4 Proposal Selection: Regional Dependence

We have investigated the regional dependence of the gender outcomes observed above. We selected the proposals depending on the location of the institute of the PI. We used Europe, North America and the rest of the World as regions in a similar manner to [68] for HST. The number of proposals from each region and the fraction that were accepted are reported in Table 4.6. It is noticeable that proposals from Europe have the highest fraction of female PIs and overall success rates. Both the proposals from Europe and North America have higher success rates for male PIs compared to female PIs. The success rate for the rest of the World is lower than elsewhere with little difference between the male and female PI success rates. This is similar to the reported by Carpenter [70] for Atacama Large Millimetre/sub-millimetre Array (ALMA).

Table 4.6 The regional dependence of proposal submission and acceptance

	No. of submitted proposals			Total	Female/Total
Region	Total	Male	Female	%	%
Europe	5277	3856	1421	49.9	26.9
North America	4467	3480	987	42.2	22.1
Rest of the world	835	661	174	7.9	20.8
	No. of accepted proposals				
	Total	Male	Female		
Europe	2270	1708	562		
North America	1764	1407	357		
Rest of the World	253	201	52		
	% Accepted proposals			% Difference	
	Total	Male	Female	Male	Female
Europe	43.0	44.3	39.5	1.3	−3.5
North America	39.5	40.4	36.3	0.9	−3.3
Rest of the world	30.3	30.4	29.9	0.1	−0.4

4.9   Discussion

We have examined how our results compare to those from missions and facilities. HST is the obvious mission for comparison as it is similarly long-lived as XMM-Newton and has an even higher average number of proposals in response to each AO. There have also been investigations of the outcomes of the observing selection processes for ESO, Canadian facilities and the National Radio Astronomy Observatory (NRAO) facilities published, which we also consider.

There have been a number of other studies that have examined the gender dependence of AO outcomes. Reid [68] reports on a study of gender-based systematic trends in the HST proposal review process for Cycles 11 through 21 (2001 to 2013). Key results from this study of the outcomes of 9400 proposals are that there appears to be a similar trend to that seen on XMM-Newton in that male PIs are more likely to succeed in achieving a successful HST proposal (23.5% success rate) than female PIs (18.1% success rate). This is a difference of 30% in favour of proposals led by male PIs. The effect is not necessarily significant in a single cycle but the imbalance is systematic with the success rate of HST female PIs consistently falling below that of their male colleagues. To help combat this effect, the Space Telescope Science Institute (STScI) adopted a system of dual-anonymous review, in which the names of the reviewers and the investigators are made known to each other only after the review has been completed. In HST Cycle 26, for the first time, proposals with female PIs had a higher success rate (8.7%) than those led by men (8.0%) [71]. However, this result is based on a single cycle which was unusual in that it had many fewer proposals than the average. Subsequent HST Cycles 27 to 29 (2019–2021) which were also conducted with dual-anonymous reviewing again showed higher male PI success rates. [72] report average male and female PI success rates of 16.2 and 14.8%, respectively for HST Cycles 26–29 which is a difference in favour of male PIs of 9.3%. The previous HST Cycles 22–25 had average male and female success rates of 25.1 and 22.2%. This is a larger difference than for Cycles 26–29 in favour of male PIs of 13.2%. Thus in HST Cycles 27 to 29 dual-anonymous reviewing probably *helped* the HST AO selection process to provide more equitable outcomes. In addition, as [72] reports, the HST success rate of early career scientists increased under dual-anonymous reviewing from ~5% to ~30% with a significant increase in the number of first-time PIs. It is unclear if this leads to a more scientifically productive mission.

The analysis presented here indicates that the XMM-Newton proposal selection process has comparable gender outcomes to that of HST after dual-anonymous reviewing was implemented by STScI. Note that we do not refer to a "bias" in the XMM-Newton proposal selection since the different success rates may result from factors not associated with the proposal selection process itself. The XMM-Newton proposal selection process is clearly well optimised, as would be expected following so many AO calls. We do not see a strong case for the implementation of dual-anonymous proposal handling for XMM-Newton. Such a change could have unexpected consequences given the complexities of the selection process.

An investigation of the time allocation process at ESO is reported in [73]. This covers an interval of 8 years and involved about 3000 PIs. Female PIs were found to have a significantly lower chance of being awarded a top rank compared to male PIs with a male/female ratio of 1.39 ± 0.05. The paper suggests that the principal explanation for the difference may be due to the average higher seniority of the male PIs assuming that more senior scientists of both genders write better proposals and thus succeed more often at obtaining ESO telescope time. However, no attempt was made to quantify if this effect can account for the observed differences. Nevertheless, [73] concludes that the ESO review process itself introduces extra gender differences.

The results of the Canadian Time Allocation Committee rankings for the Canadian share of the Canada-France-Hawaii Telescope (CFHT) and Gemini Observatory time have been investigated by Spekkens et al. [74]. They find that proposals led by males are more likely to obtain higher scores than those led by females. This difference was present for both proposers who had and did not have faculty positions implying that seniority was not the cause. An investigation of the gender-related systematics in the proposal review processes for the four facilities operated by NRAO: the Jansky Very Large Array (JVLA), the Very Long Baseline Array (VLBA), the Green Bank Telescope (GBT) and ALMA are reported in [75] and [70]. Similarly to HST, before the introduction of dual-anonymous reviewing, and similarly as well to the ESO study there are significant gender-related effects in the proposal rankings in favour of male PIs compared to female PIs for all four facilities with varying degrees of confidence reflecting the different number of proposals.

In summary, the difference observed in the XMM-Newton proposal selection process of 5–15% in favour of male PIs is smaller than those reported for some ESO, and NRAO facilities [73, 75] and for HST prior to the implementation of dual-

Table 4.7 Summary of XMM-Newton proposal acceptance percentages for AO-2 to AO-21 for both proposal numbers and awarded time. The overall male/female difference covers the range of given male/female ratios and is expressed as a percentage

Priority	Parameter	All	Proposal PI Male	Female
	Proposals Submitted	10,579	7997	2582
A,B,C	Number Accepted	4287	3316	971
	Percentage Accepted	40.5%	41.5	37.7
	Ratio Male/Female		1.100	
A,B	Number Accepted	2467	1929	538
	Percentage Accepted	23.3	24.1	20.9
	Ratio Male/Female		1.155	
	Time Requested (s)	1.97×10^9	1.49×10^9	4.81×10^8
A,B,C	Time Accepted (s)	4.82×10^8	3.68×10^8	1.14×10^8
	Percentage time accepted	24.4%	24.7%	23.6%
	Ratio male/female		1.045	
A,B	Time accepted(s)	2.78×10^8	2.17×10^8	6.08×10^7
	Percentage time accepted	14.1%	14.6%	12.6%
	Ratio male/female		1.154	
	Overall male/female difference		5–15%	

anonymous reviewing [68]. Once dual-anonymous reviewing was implemented for HST the male and female PI proposal acceptance difference falls to 9.3% [72], comparable to that seen here with XMM-Newton of 5–15% (Table 4.7). The success rate of scientists prior to obtaining their PhDs of ~35% on XMM-Newton is comparable to that on HST following dual-anonymous reviewing. This suggests that the XMM-Newton proposal selection processes produces outcomes consistent with, or at least close to, the best available science. It does not however provide gender parity and a number of activities are proposed below to investigate this matter.

4.10   Further Activities

In order to better understand some of the issues arising from the analysis of the XMM-Newton proposal selection processes, the following would be useful:

1. An investigation of the uncertainties in the proposal selection process. This could be done by comparing the results of different representative OTACs evaluating the same set of proposals. Given the amount of work involved in an OTAC evaluation this would be a major undertaking. Instead, it may be possible to gain insights into the process by looking at how proposals that are evaluated by multiple panels are assessed.

2. A study of the impact of native language on the success rate of proposals. Non-native language is increasingly being recognised as a major barrier in science [76] implying that native English speakers are more successful than their counterparts in obtaining observing time. This does not appear to be the case for XMM-Newton as countries which are expected to have large fractions of native English speakers, such as the United Kingdom and the United States, do not have unusually high success rates for their XMM-Newton proposals (see Table 4.2).

3. A further investigation of the acceptance rates of male and female proposers for different geographical regions – for example dividing Europe into Northern and Southern regions. This could show that regional differences play a role and provide insights into the causes of the different success rates.

4. An investigation into the publication rates of early and late career scientists who have been awarded XMM-Newton observing time. Are late career scientists better at exploiting their observations and so produce relatively more publications and citations than their early career colleagues? More generally, what is the effect on the science return of a mission if the observations are dominated by late or early career scientists. This is a complex issue as e.g., a late career scientist may have made substantial contributions to a proposal from an early career colleague, so improving its chances of success.

Acknowledgments We thank Pedro Rodriquez and the XMM-Newton SOC staff. Fred Jansen is acknowledged for his efforts in astro-archeology. We thank Neill Reid at the STScI who provided the results of the most recent HST observing cycles. This research was supported by the International Space Science Institute (ISSI) in Bern, through the ISSI Working Group project 'The Scientific Performance of ESA's Science Missions'.

References

1. F. Jansen, D. Lumb, B. Altieri, et al., XMM-Newton observatory. I. The spacecraft and operations. Astron. Astrophys. **365**, L1–L6 (2001)
2. L. Strüder, U. Briel, K. Dennerl, et al., The European photon imaging camera on XMM-Newton: the pn-CCD camera. Astron. Astrophys. **365**, L18–L26 (2001)
3. J.W. den Herder, A.C. Brinkman, S.M. Kahn, et al., The reflection grating spectrometer on board XMM-Newton. Astron. Astrophys. **365**, L7–L17 (2001)
4. K.O. Mason, A. Breeveld, R. Much, et al., The XMM-Newton optical/UV monitor telescope. Astron. Astrophys. **365**, L36–L44 (2001)
5. M.J.L. Turner, A. Abbey, M. Arnaud, et al., The European photon imaging camera on XMM-Newton: the MOS cameras: the MOS cameras. Astron. Astrophys. **365**, L27–L35 (2001)
6. M.C. Weisskopf, The advanced X-ray astrophysics facility, in *IAU Colloq. 123: Observatories in Earth Orbit and Beyond*, ed. by Y. Kondo. Astrophysics and Space Science Library, vol. 166 (1990), p. 71
7. H.K. Connor, J.A. Carter, Exospheric neutral hydrogen density at the nominal 10 RE subsolar point deduced from XMM-Newton X-ray observations. J. Geophys. Res. **124**(3), 1612–1624 (2019)
8. P. Medvedev, M. Gilfanov, S. Sazonov, et al., XMM-Newton observations of the extremely X-ray luminous quasar CFHQS J142952+544717=SRGE J142952.1 + 544716 at redshift z = 6.18. Month. Not. R. Astron. Soc. **504**(1), 576–582 (2021)

9. M.J. Page, C. Simpson, D.J. Mortlock, et al., X-rays from the redshift 7.1 quasar ULAS J1120+0641. Month. Not. R. Astron. Soc. **440**, L91–L95 (2014)

10. A. Moretti, L. Ballo, V. Braito, et al., X-ray observation of ULAS J1120+0641, the most distant quasar at z = 7.08. Astron. Astrophys. **563**, A46 (2014)

11. D. Farrah, R. Priddey, R. Wilman, et al. The X-Ray spectrum of the z=6.30 QSO SDSS J1030+0524. Astrophys. J. Lett. **611**(1), L13–L16 (2004)

12. M. Santos-Lleo, N. Schartel, H. Tananbaum, et al., The first decade of science with Chandra and XMM-Newton. Nature **462**(7276), 997–1004 (2009)

13. B.J. Wilkes, W. Tucker, N. Schartel, et al., X-ray astronomy comes of age. Nature **606**(7913), 261–271 (2022)

14. K. Dennerl, C.M. Lisse, A. Bhardwaj, et al., First observation of Mars with XMM- Newton. High resolution X-ray spectroscopy with RGS. Astron. Astrophys. **451**(2), 709–722 (2006)

15. W.R. Dunn, G. Branduardi-Raymont, L.C. Ray, et al., The independent pulsations of Jupiter's northern and southern X-ray auroras. Nat. Astron. **1**, 758–764 (2017)

16. S.J. Bolton, J. Lunine, D. Stevenson, et al., The juno mission. Space Sci. Rev. **213**(1–4), 5–37 (2017)

17. Y. Zhonghua, W.R. Dunn, E.E. Woodfield, et al., Revealing the source of Jupiter's x-ray auroral flares. Sci. Adv. **7**(28), eabf0851 (2021)

18. R. Di Stefano, J. Berndtsson, R. Urquhart, et al., A possible planet candidate in an external galaxy detected through X-ray transit. Nat. Astron. **5**, 1297–1307 (2021)

19. I. Pillitteri, S.J. Wolk, J. Lopez-Santiago, et al., The Corona of HD 189733 and its X-Ray activity. Astrophys. J. **785**(2), 145 (2014)

20. K. Poppenhaeger, Helium absorption in exoplanet atmospheres is connected to stellar coronal abundances. Month. Not. Royal Astron. Soc. **512**(2), 1751–1764 (2022)

21. A. Papitto, C. Ferrigno, E. Bozzo, et al., Swings between rotation and accretion power in a binary millisecond pulsar. Nature **501**(7468), 517–520 (2013)

22. N. Rea, P. Esposito, R. Turolla, et al., A low-magnetic-field soft gamma repeater. Science **330**(6006), 944 (2010)

23. A. Tiengo, P. Esposito, S. Mereghetti, et al., A variable absorption feature in the X-ray spectrum of a magnetar. Nature **500**(7462), 312–314 (2013)

24. G.L. Israel, A. Belfiore, L. Stella, et al., An accreting pulsar with extreme properties drives an ultraluminous x-ray source in NGC 5907. Science **355**(6327), 817–819 (2017)

25. F. Furst, D.J. Walton, F.A. Harrison, et al., Discovery of coherent pulsations from the ultraluminous X-Ray source NGC 7793 P13. Astrophys. J. Letter **831**(2), L14 (2016)

26. A.C. Fabian, A. Zoghbi, R.R. Ross, et al., Broad line emission from iron K-and L-shell transitions in the active galaxy 1H0707-495. Nature **459**(7246), 540–542 (2009)

27. G. Risaliti, F.A. Harrison, K.K. Madsen, et al., A rapidly spinning supermassive black hole at the centre of NGC 1365. Nature **494**(7438), 449–451 (2013)

28. W.N. Alston, A.C. Fabian, E. Kara, et al., A dynamic black hole corona in an active galaxy through X-ray reverberation mapping. Nat. Astron. **4**, 597–602 (2020)

29. D.R. Wilkins, L.C. Gallo, E. Costantini, et al., Light bending and X-ray echoes from behind a supermassive black hole. Nature **595**(7869), 657–660 (2021)

30. F. Tombesi, M. Cappi, J.N. Reeves, et al., Evidence for ultra-fast outflows in radio-quiet AGNs. I. Detection and statistical incidence of Fe K-shell absorption lines. Astron. Astrophys. **521**, A57 (2010)

31. F. Tombesi, M. Cappi, J.N. Reeves, et al., Evidence for ultra-fast outflows in radio-quiet active galactic nuclei. II. Detailed photoionization modeling of Fe K-shell absorption lines. Astrophys. J. **742**(1), 44 (2011)

32. F. Tombesi, M. Cappi, J.N. Reeves, et al., Evidence for ultrafast outflows in radio-quiet AGNs - III. Location and energetics. Month. Not. R. Astron. Soc. **422**(1), L1–L5 (2012)

33. E. Nardini, J.N. Reeves, J. Gofford, et al., Black hole feedback in the luminous quasar PDS 456. Science **347**(6224), 860–863 (2015)

34. M.L. Parker, C. Pinto, A.C. Fabian, et al., The response of relativistic outflowing gas to the inner accretion disk of a black hole. Nature **543**(7643), 83–86 (2017)

35. S. Komossa, J. Halpern, N. Schartel, et al., A huge drop in the X-Ray luminosity of the nonactive galaxy RX J1242.6-1119A, and the first postflare spectrum: testing the tidal disruption scenario. Astrophys. J. Lett. **603**(1), L17–L20 (2004)

36. D. Lin, J. Strader, E.R. Carrasco, et al., A luminous X-ray outburst from an intermediate-mass black hole in an off-centre star cluster. Nat. Astron. **2**, 656–661 (2018)

37. D. Lin, J. Guillochon, S. Komossa, et al., A likely decade-long sustained tidal disruption event. Nat. Astron. **1**, 0033 (2017)

38. D.R. Pasham, R.A. Remillard, P.C. Fragile, et al., A loud quasi-periodic oscillation after a star is disrupted by a massive black hole. Science **363**(6426), 531–534 (2019)

39. R.C. Reis, J.M. Miller, M.T. Reynolds, et al., A 200-second quasi-periodicity after the tidal disruption of a star by a dormant black hole. Science **337**(6097), 949 (2012)

40. J.M. Miller, J.S. Kaastra, M.C. Miller, et al., Flows of X-ray gas reveal the disruption of a star by a massive black hole. Nature **526**(7574), 542–545 (2015)

41. E. Kara, J.M. Miller, C. Reynolds, et al., Relativistic reverberation in the accretion flow of a tidal disruption event. Nature **535**(7612), 388–390 (2016)

42. G. Miniutti, R.D. Saxton, M. Giustini, et al., Nine-hour X-ray quasi-periodic eruptions from a low-mass black hole galactic nucleus. Nature **573**(7774), 381–384 (2019)

43. R. Arcodia, A. Merloni, K. Nandra, et al., X-ray quasi-periodic eruptions from two previously quiescent galaxies. Nature **592**(7856), 704–707 (2021)

44. E. Pointecouteau, M. Arnaud, G.W. Pratt, The structural and scaling properties of nearby galaxy clusters. I. The universal mass profile. Astron. Astrophys. **435**(1), 1–7 (2005)

45. N. Werner, J. de Plaa, J.S. Kaastra, et al., XMM-Newton spectroscopy of the cluster of galaxies 2A 0335+096. Astron. Astrophys. **449**(2), 475–491 (2006)

46. J. de Plaa, N. Werner, A.M. Bykov, et al., Chemical evolution in Sersic 15903 observed with XMM-Newton. Astron. Astrophys. **452**(2), 397–412 (2006)

47. J.R. Peterson, F.B.S. Paerels, J.S. Kaastra, et al., X-ray imaging-spectroscopy of Abell 1835. Astron. Astrophys. **365**, L104–L109 (2001)

48. T. Tamura, J.S. Kaastra, J.R. Peterson, et al., X-ray spectroscopy of the cluster of galaxies Abell 1795 with XMM-Newton. Astron. Astrophys. **365**, L87–L92 (2001)

49. J.S. Kaastra, C. Ferrigno, T. Tamura, et al., XMM-Newton observations of the cluster of galaxies Sérsic 159-03. Astron. Astrophys. **365**, L99–L103 (2001)

50. J.S. Sanders, A.C. Fabian, R.K. Smith, et al., A direct limit on the turbulent velocity of the intracluster medium in the core of Abell 1835 from XMM-Newton. Month. Not. R. Astron. Soc. **402**(1), L11–L15 (2010)

51. C. Pinto, J.S. Sanders, N. Werner, et al., Chemical enrichment RGS cluster sample (CHEERS): constraints on turbulence. Astron. Astrophys. **575**, A38 (2015)

52. F. Nicastro, J. Kaastra, Y. Krongold, et al., Observations of the missing baryons in the warm-hot intergalactic medium. Nature **558**(7710), 406–409 (2018)

53. R. Massey, J. Rhodes, R. Ellis, et al., Dark matter maps reveal cosmic scaffolding. Nature **445**(7125), 286–290 (2007)

54. E. Bulbul, M. Markevitch, A. Foster, et al., Detection of an unidentified emission line in the stacked X-Ray spectrum of galaxy clusters. Astrophys. J. **789**(1), 13 (2014)

55. A. Boyarsky, O. Ruchayskiy, D. Iakubovskyi, et al., Unidentified line in X-Ray spectra of the andromeda galaxy and perseus galaxy cluster. Phys. Rev. Lett. **113**(25), 251301 (2014)

56. C. Dessert, N.L. Rodd, B.R. Safdi, The dark matter interpretation of the 3.5-keV line is inconsistent with blank-sky observations. Science **367**(6485), 1465–1467 (2020)

57. M. Pierre, F. Pacaud, C. Adami, et al., The XXL survey. I. Scientific motivations - XMM-Newton observing plan - follow-up observations and simulation programme. Astron. Astrophys. **592**, A1 (2016)

58. F. Pacaud, N. Clerc, P.A. Giles, et al., The XXL survey. II. The bright cluster sample: catalogue and luminosity function. Astron. Astrophys. **592**, A2 (2016)

59. M. Arnaud, Evolution of clusters and cosmology. Astron. Nachr. **338**(342), 342–348 (2017)

60. G. Risaliti, E. Lusso, Cosmological constraints from the hubble diagram of quasars at high redshifts. Nat. Astron. **3**, 272–277 (2019)

61. J.U. Ness, A.N. Parmar, L.A. Valencic, et al., XMM-Newton publication statistics. Astron. Nachr. **335**(2), 210 (2014)
62. A.H. Rots, S.L. Winkelman, G.E. Becker, Chandra publication statistics. Publ. Astr. Soc. Pac. **124**(914), 391 (2012)
63. D. Apai, J. Lagerstrom, I.N. Reid, et al., Lessons from a high-impact observatory: the hubble space telescope's science productivity between 1998 and 2008. Publ. Astr. Soc. Pac. **122**(893), 808 (2010)
64. V. Trimble, P. Zaich, T. Bosler, Productivity and impact of space-based astronomical facilities. Publ. Astr. Soc. Pac. **118**(842), 651–655 (2006)
65. V. Trimble, J.A. Ceja, Productivity and impact of astronomical facilities: a statistical study of publications and citations. Astron. Nachr. **328**(9), 983–994 (2007)
66. V. Trimble, J.A. Ceja, Productivity and impact of astronomical facilities: three years of publications and citation rates. Astron. Nachr. **329**(6), 632–647 (2008)
67. V. Trimble, Telescopes in the mirror of scientometrics. Exp. Astron. **26**(1–3), 133–147 (2009)
68. I.N. Reid, Gender-correlated systematics in HST proposal selection. Publ. Astron. Soc. Pac. **126**(944), 923 (2014)
69. G.J. Babu, E.D. Feigelson, Goodness-of-fit and all that! ASP Conf. **351**, 127 (2006)
70. J. Carpenter, Systematics in the ALMA proposal review rankings. Publ. Astron. Soc. Pac. **132**(1008), 024503 (2020)
71. L. Strolger, N. Priyamavada, Doling out Hubble time with dual-anonymous evaluation. Phys. Today Comment. 1 March 2019. https://doi.org/10.10163/PT.6.3.20190301a (2019)
72. N. Reid, Personal communication (2023)
73. F. Patat, Gender systematics in telescope time allocation at ESO. The Messenger **165**, 2–9 (2016)
74. K. Spekkens, N. Cofie, D. Crabtree, Sex-disaggregated systematics in Canadian time allocation committee telescope proposal reviews, in *Observatory Operations: Strategies, Processes, and Systems VII*. Society of Photo-Optical Instrumentation Engineers (SPIE) Conference Series, vol. 10704 (2018). p. 107040L
75. C.J. Lonsdale, F.R. Schwab, G. Hunt, Gender-related systematics in the NRAO and ALMA proposal review processes (2016). arXiv e-prints
76. M. Lenharo, The true cost of science's language barrier for non-native English speakers. Nature **619**, 678–679 (2023)

Chapter 5
INTEGRAL Observing Time Proposals

Erik Kuulkers, Celia Sánchez-Fernández, and Arvind Parmar

Abstract We examine the outcomes of the regular announcements of observing opportunities for ESA's gamma-ray observatory INTEGRAL issued between 2000 and 2021. We investigate how success rates vary with the lead proposer's gender, "academic age" and the country where the proposer's institute is located. The more than 20 years operational lifetime enable the evolution of the community proposing for INTEGRAL to be probed. We determine proposal success rates for high-priority and all proposals using both the numbers of accepted proposals and the amounts of awarded observing time. We find that male lead proposers are between 2–11% more successful than their female counterparts in obtaining INTEGRAL observations. We investigate potential correlations between the female-led proposal success rates and the amount of female participation in the Time Allocation Committee.

Keywords INTEGRAL · Science results · Science productivity · Announcement of observing opportunities · Over-subscription · Open time · Peer review · Time Allocation Committee · Joint programmes · Panel members · Panel chairs · Observing priorities · Gender balance · User group · Proposers' institutes countries · Academic age · Proposal acceptance rates · Gender · Regional dependence · Dual-anonymous reviews · Further activities

E. Kuulkers (✉)
Directorate of Science, ESA, ESTEC, Noordwijk, The Netherlands
e-mail: Erik.kuulkers@esa.int

C. Sánchez-Fernández
Directorate of Science, ESA, ESAC, Villanueva de la Cañada, Madrid, Spain
e-mail: celia.sanchez@ext.esa.int

A. Parmar
Department of Space and Climate Physics, MSSL/UCL, Dorking, UK
e-mail: arvind.parmar@ucl.ac.uk

© The Author(s) 2025
A. Parmar et al., *ESA Science Programme Missions*, ISSI Scientific Report Series 18,
https://doi.org/10.1007/978-3-031-69004-4_5

5.1 Introduction

INTErnational Gamma-Ray Astrophysics Laboratory (INTEGRAL) was launched from the Baikonur cosmodrome on 17 October 2002 into a 72-hour highly elliptical orbit by a Proton rocket. The nominal mission duration was 2 years with resources available for a longer extended mission. In January 2015 the orbit was changed to a 64 hour orbital period to ensure the safe disposal of the mission in 2029. This ESA-led mission includes contributions from Russian Federation (launcher) and National Aeronautics and Space Administration (NASA) (Goldstone ground station). INTEGRAL provides an unprecedented combination of celestial imaging and spectroscopy over a wide range of X-ray and gamma-ray energies as well as simultaneous optical monitoring [1, 2]. Many details about the INTEGRAL spacecraft, orbit, instruments, scientific aims and first results can be found in volume 411 of A&A (2003), a special Astronomy & Astrophysics issue dedicated to INTEGRAL. The INTEGRAL payload consists of two gamma-ray instruments, one of which is optimised for 15 Kilo Electron Volt (keV) to 10 Million Electron Volt (MeV) high-resolution imaging (Imager On-Board the INTEGRAL Spacecraft (IBIS); [3]) and the other for 20 keV to 8 MeV high-resolution spectroscopy (Spectrometer on INTEGRAL (SPI); [4]). IBIS provides an angular resolution of 12' Full-Width at Half Maximum (FWHM) over a 30×30 degrees^2 Field of View (FOV) and an energy resolution, $E/\Delta E$, of ~ 12 FWHM at 100 keV. SPI provides an angular resolution of 2.7 degrees FWHM over a 30×30 degrees^2 FOV and an $E/\Delta E$ of ~ 500 FWHM at 1.3 MeV. Both instruments provide millisecond time resolution. The extremely broad energy range of IBIS is covered by two separate detector arrays, INTEGRAL Soft Gamma-Ray Imager (ISGRI) (15–500 keV) and Pixellated Imaging Caesium Iodide Telescope (PICsIT) (0.2–10 MeV). The payload is completed by two Joint European X-ray Monitor (JEM-X); 3–35 keV; [5]) and optical (Optical Monitor Camera (OMC); V-band; [6]) monitors. The instruments are co-aligned and are normally operated simultaneously. By using the anti-coincidence shields of IBIS and SPI as active detectors, an omni-directional sensitivity above 75 keV is achieved. The highly eccentric orbit (period 2.7-days) allows the full sky to be continuously monitored for about 85% of the time (when INTEGRAL is above the Earth's radiation belts). The fraction of sky occulted by the Earth is much smaller than for satellites in low-Earth orbits.

The Mission Operations Centre (MOC) is located at European Space Operations Centre (ESOC), in Darmstadt, Germany. The Science Operations Centre (SOC) was originally in European Science and Technology Centre (ESTEC), Noordwijk, The Netherlands but moved to European Space Astronomy Centre (ESAC), near Madrid, Spain in February 2005. The SOC receives observation proposals and optimises the accepted ones into an observation plan consisting of a time line of target pointings together with the corresponding instrument configurations. The INTEGRAL Science Data Centre (ISDC) receives the science telemetry plus the relevant ancillary spacecraft data from the MOC which is responsible for the operations of the spacecraft and payload. The ISDC Data Centre for Astrophysics [7] processes the received telemetry and generates standard data products which are distributed and archived. The archive at the ISDC receives about 300 independent visits per month, a rate that has been stable over the past decade. INTEGRAL data can be processed using the Off-line Scientific Analysis (OSA) software provided by the ISDC. This includes pipelines for the reduction of INTEGRAL data from all four instruments. The three high-energy instruments use coded masks to provide imaging information. This means that photons from a source within the FOV are distributed over the detector area in a pattern determined by the position of the source in the FOV. Source positions and intensities are determined by matching the observed distribution of counts with those produced by the mask modulation. The energy range and use of coded masks means that observation lengths are usually longer than for directly imaging instruments due to the higher background in the source extraction regions.

5.2 Science Results

INTEGRAL has become part of the newly-established multi-messenger astro-physics science network, measuring gamma-rays coincident with Gravitational Wave (GW), ultra-high energy neutrino signals and Fast Radio Bursts (FRBs), so linking the new non-electromagnetic domains with the hard X- and gamma-ray electromagnetic radiation [8–10]. This synergy is achieved primarily by exploiting two of INTEGRAL's key capabilities—the highly efficient coverage of nearly the whole sky and the rapid reaction capability for unexpected events. INTEGRAL can rapidly (typically within half a day) re-point and conduct Target of Opportunity (ToO) observations. Of all the ToO observations performed until the end of 2021, 4% started within 8 hours, while 85% started 3 days or later after the alert.

A key part of INTEGRAL's legacy is the continuing discovery of new high-energy sources, now amounting to more than 700 [11, 12]. Using multi-wavelength follow-up observations, about 250 of the newly discovered sources have been firmly identified, and their dominant emission mechanisms recognised. The classes of objects discovered include Seyfert galaxies, blazars, cataclysmic variables, supergiant fast X-ray transients. INTEGRAL also provides unique measurements of the decay lines of radioactive isotopes, e.g., from extragalactic supernovae and for the study of stellar feedback in our Galaxy. The extent and morphology of the enigmatic positron annihilation in the central regions of the Galaxy, possibly linked to dark-matter decay, as well as of ^{26}Al and ^{60}Fe continues to be mapped by INTEGRAL [13–15]. INTEGRAL is also performing pioneering and unique gamma-ray polarization studies of Gamma-ray Burst (GRB)s and bright Galactic sources, which are fundamental in understanding the emission mechanisms. The rich science which has resulted from INTEGRAL's observations over its lifetime are presented in the various review articles in the special issue of New Astronomy Reviews "Fifteen-plus years of INTEGRAL science" available at

https://www.sciencedirect.com/journal/new-astronomy-reviews/special-issue/
10K9PV78WN4. Here we provide some examples to illustrate the scientific impact
of INTEGRAL.

Gravitational Wave Sources The INTEGRAL and Fermi [16] detection of the
short GRB (GRB 170817A) linked to a GW event (GW170817) on 17 August 2017
caused by the merger of a binary neutron star [17, 18], was a fundamental milestone
for the wider astrophysics and physics communities. Using INTEGRAL, the Fermi
localisation area of the GRB could be improved by more than a factor two; this has
been important to prove, not only the temporal, but also the spatial coincidence of
the gamma-ray emission and the GW event. INTEGRAL has provided instantaneous
and complete coverage of the large localisation regions of most of the other GW
events detected by the LIGO/Virgo collaboration. The majority of them have been
associated with binary black holes, for which most models do not predict detectable
electromagnetic signals: whenever INTEGRAL was above the radiation belts, it
provided strict upper limits.

Bright Transients Several bright black-hole binary transients have led to unique,
extensive ToO campaigns with INTEGRAL. In 2015, V404 Cygni, one of the closest
(2.4 kilo parsec (kpc)) and best-established accreting black-hole (about 9 $M_\odot$)
binaries, became the brightest object in the X-ray sky (up to 50 times the flux
of the Crab). Its outburst showed highly unusual behaviour with repeated bright
optical and X-rays flashes on time scales shorter than one hour. It was observed
almost continuously by INTEGRAL simultaneously in the optical and hard X-
ray bands, which was not possible with any other mission. More than 40 refereed
publications have been written using these data. Transients of this type can provide
unique diagnostics of the physical processes operating near black holes through
the study of optical and hard X-ray to gamma-ray emission [19–29], including
positron annihilation signals [26, 30–32], and gamma-ray polarisation [33]. The
sensitive study of Comptonisation, pair emission, and polarisation above $\sim$100 keV
are unique strengths of INTEGRAL.

Supernovae INTEGRAL discovered gamma-ray emission from the decay of
^{56}Ni and ^{56}Co in a Type Ia supernova, SN2014J. While confirming theoretical
expectations these measurements significantly added new constraints on how ^{56}Ni
ejecta are spatially and kinematically distributed throughout an expanding envelope
[34–39]. For core-collapse supernovae, INTEGRAL spectroscopy of lines from the
decay of ^{44}Ti likewise provided essential kinematic information from the deep
interiors of such supernovae (Cas A; [40, 41]), which are otherwise inaccessible.
Both families of explosions have now been recognised to be very diverse. Nuclear,
as well as non-thermal, emission occurs within INTEGRAL's energy range allowing
for essential follow-up observations.

Fast Radio Bursts One of the currently intriguing mysteries is the origin of
millisecond duration, 10^{40} erg, powerful FRBs. The majority of FRBs do not
repeat, but a growing number of newly discovered events (now >10) are repeaters
[42–44]. In FRB180916.J0158+65 a 16-day periodicity suggests an origin from a

compact object either in a binary system, or in an isolated, precessing, magnetar [43]. INTEGRAL is well suited to study such systems with typical emission at high energies [45, 46]. The near all-sky monitoring capability of INTEGRAL allowed a magnetar flare associated with a Galactic FRBs in the soft gamma-ray repeater SGR 1935+2154 to be found, showing that some FRBs are associated with magnetars [9]. Since FRBs are extragalactic, this event provides a "missing link" to a population of extragalactic magnetars. GRB200415A, a giant flare from a magnetar in NGC 253 [47] may be such a source.

Other Transients ToO observations based on multi-wavelength studies are important for understanding the diversity of transient events. A recent example is AT2018cow (a supernova variant), also referred to as a fast-rising blue optical transient, which revealed its internal engine as a temporary hard X-ray spectral component detected by IBIS up to about 100 keV, in addition to luminous and highly variable soft X-rays [48]. Another case of a rapidly variable and rising nuclear transient with soft gamma-ray emission was AT2019pev, which may be evidence of a new class of flares from accreting supermassive black holes [49, 50]. This opens a still unexplored window bridging optical transients to explosive events, never observed before in this phase.

Serendipitous Science The INTEGRAL Radiation Environment Monitor (IREM) and a few of the Radiation-sensitive Field-Effect Transistors (RadFETs) on IBIS are providing important information about the Earth's local environment and contributing to interplanetary space weather [51, 52]. IREM is also used as a calibrator of similar radiation monitors onboard other spacecrafts [53]. Another area where INTEGRAL provides serendipitous science includes Solar System research, with the study of Earth's aurorae and Solar flares [54, 55].

5.3 Scientific Productivity

Up to the end of 2021, there have been close to 2000 refereed publications from INTEGRAL that meet the inclusion criteria given in Sect. 2.2. The scientific productivity of INTEGRAL is screened monthly by the INTEGRAL SOC, following an approach consistent with those used by the XMM-Newton [56], Chandra [57] and Hubble Space Telescope (HST) [58] projects. The resulting data are stored in a dedicated database from which information can be extracted when needed. The INTEGRAL SOC determines the Observation Identifier (ObsID) and instruments used to provide the scientific results reported in each publication. The ObsIDs are then used to access the INTEGRAL Science Legacy Archive (ISLA) to provide detailed information on the observations and proposal leading to relevant data set. The information obtained is then combined to allow the scientific productivity of the mission to be investigated. The major conclusions of this work are the following:

- An average of 51 scientists per year become lead authors for the first time on a refereed paper which directly uses INTEGRAL data.

- Each refereed INTEGRAL publication receives an average of around five citations per year with a long-term citation rate of two citations per year, more than five years after publication.
- About eighty percent of the articles citing INTEGRAL articles are not primarily INTEGRAL observational papers.
- The distribution of elapsed time between the execution of the latest observation used in a publication and the date of that publication peaks at about 2.5 years. The publication of observations taken under the ToO program (see below) shows a similar distribution. In other words, no significant differences are seen by the publication of INTEGRAL Open Time and ToO observations.
- At least 95% of the science observing time is used in at least one publication after 7 elapsed years.
- The scientific productivity of INTEGRAL measured by the publication rate, number of new authors and citation rate, has fluctuated over the time, and experienced a significant boost in 2017, following the INTEGRAL contribution to the detection of the electro-magnetic counterpart to GW170817.

5.4   Observing Time

INTEGRAL is an observatory executing observations mainly selected through peer review. Usually there are annual Announcement of Opportunity (AO)s, see Table 5.1. The first announcement (AO-1) was issued in November 2000, prior to launch, and received 277 proposals with a very high over-subscription in time of a factor 19.7 (see Sect. 5.5 and Table 5.1). AO-2 was issued in July 2003 and covered observations for a one year interval. Due to the impending move of the INTEGRAL Science Operations Centre (ISOC) from ESTEC to ESAC in 2005, AO-3 covered an interval of 18 months and was issued in September 2004. Subsequent AOs, except for AO-6 and AO-7, covered intervals of one year and were issued in the Spring of each year. AO-6 was initially foreseen to cover 1 year, but in November 2008 it was decided to extend the cycle by two months. AO-7 also covered an interval of 14 months. The following types of observing time are available:

- Normal observations: observations which do not have special scheduling requirements allowing for the most efficient scheduling. Observation durations are <1 Millions of Seconds (Msec).
- Key Programme observations: Observations intended to carry out scientific investigations requiring a significant amount of the total non-Target-of-Opportunity (non-ToO) observing time during an AO cycle (>1 Msec), but also accommodating various scientific aims. These cannot be ToO observations.
- Fixed Time observations: observations with special scheduling requirements such as phase-dependent observations of a binary system, coordinated multi-wavelength observations, or a sequence of observations separated by a time

Table 5.1 INTEGRAL AO summary. The numbers of TAC members include the panel chairs

AO	Year issued	Observation interval	Total no. of proposals	Over-subs. (time)	TAC Members Male	TAC Members Female	TAC Panel chairs Male	TAC Panel chairs Female
1	2000	30 Dec 2002–16 Dec 2003	277	19.7	22	5	4	0
2	2003	17 Dec 2003–17 Feb 2005	151	7.9	21	6	4	0
3	2004	18 Feb 2005–15 Aug 2006	112	5.7	20	6	3	0
4	2006	16 Aug 2006–16 Aug 2007	145	7.7	15	4	3	0
5	2007	17 Aug 2007–16 Aug 2008	183	6.7	17	2	3	0
6	2008	17 Aug 2008–15 Oct 2009	184	5.3	19	0	3	0
7	2009	16 Oct 2009–31 Dec 2010	157	3.8	13	3	3	0
8	2010	01 Jan 2011–31 Dec 2011	132	4.1	12	4	3	0
9	2011	01 Jan 2012–31 Dec 2012	101	2.6	13	3	2	1
10	2012	01 Jan 2013–31 Dec 2013	97	2.9	13	3	2	1
11	2013	01 Jan 2014–31 Dec 2014	87	2.1	14	2	1	2
12	2014	01 Jan 2015–31 Dec 2015	77	4.5	13	3	1	2
13	2015	01 Jan 2016–31 Dec 2016	63	4.4	12	4	3	0
14	2016	01 Jan 2017–31 Dec 2017	58	5.2	12	4	3	0
15	2017	01 Jan 2018–31 Dec 2018	65	3.9	11	5	2	1
16	2018	01 Jan 2019–31 Dec 2019	62	5.2	11	5	2	1
17	2019	01 Jan 2020–31 Dec 2020	63	3.8	12	4	1	2
18	2020	01 Jan 2021–31 Dec 2021	52	3.5	11	5	1	2
19	2021	01 Jan 2022–31 Dec 2022	49	3.1	11	5	2	1
Total			**2115**		**272**	**73**	**46**	**13**

interval. Such observations usually reduce scheduling efficiency, because the spacecraft must be pointing towards a particular source at a particular time.

- ToO observations that have critical scheduling requirements and are meant as a fast response to new phenomena, such as X-ray novae outbursts, flares from active galactic nuclei, supernovae, and bright states of galactic micro-quasars. ToOs can be for known or unknown sources identified by their class.
- Multi-year proposals spanning two annual AOs for "Key Programmes" requiring substantial amounts of observing time that cannot be accommodated in a single AO.
- Joint observations with the XMM-Newton, Swift, NuSTAR, and Fermi missions.
- Guaranteed Time: from 2002 to 2008 the observing programme was divided into Open Time for the general observer (General Programme (GP)) and the Guaranteed Time (GT) for the INTEGRAL Science Working Team (ISWT) (Core Programme (CP)). The GT during the CP was the return to the ISWT for their contributions to the development and execution of the INTEGRAL project. The CP decreased from 35% in the first year to 20% in the sixth year after launch; from 2009 there was no further CP (see Table 5.2). From AO-1 to AO-6 27% of the Open Time was reserved for scientists from the Russian Federation in return for the provision of the launcher. From AO-7 this reduced to 25%. The TAC awards this guaranteed return on the same basis of scientific merit as all other proposals.
- Discretionary Time: The INTEGRAL Project Scientist can grant "Discretionary Time" which is ~5% of the available observing time. This can be used for unanticipated ToOs—observations where a timely observation outside the normal AO cycle is likely to result in a significant scientific impact. Discretionary Time observations may have a proprietary period of six months, or may be made publicly available as soon as the relevant data files have been created.
- Around 5% of the available observing time is used to maintain the calibration and monitor instrument health.

The detailed contents of the AOs varied. In AO-4 there was a pilot Key Programme for proposals asking for >1 Msec of observing time. In AO-5 and AO-6 there were dedicated calls for Key Programmes which could continue for more than one AO. In AO-7 to AO-19 Key Programme proposals which lasted for up to 2 years could be submitted. If accepted, these had to be resubmitted the following year. In AO-5 Key Programme proposals accepted for more than one year were

Table 5.2 INTEGRAL Guaranteed Time (GT) compared to the available observing time

Period	GT (%)	Mission phase
17 Dec 2002–16 Dec 2003	35	Nominal mission
17 Dec 2003–16 Dec 2004	30	Nominal mission
17 Dec 2004–16 Dec 2005	25	Extended mission
17 Dec 2005–16 Dec 2006	25	Extended mission
17 Dec 2006–16 Dec 2007	25	Extended mission
17 Dec 2007–16 Dec 2008	20	Extended mission

automatically taken over to the next year. From AO-4 to AO-11 there were separate calls for data rights proposals. These were for targets in the large IBIS and SPI FOVs that were observed "serendipitously" in other proposals. Accepted proposals were awarded zero observing time but had the right to analyse data from the approved target, or targets. A few of the proposers in AO-1 also asked for serendipitous time in other programmes, and these were also treated as data rights proposals. From AO-12 to AO-19 the two types of proposals were merged, and data right proposers had to request time, but once accepted, they were awarded zero time. Data from fast transient events (such as GRBs) may be contained in the science data of the operating instruments. Data on such an event itself can be asked for, and are handled, in principle, as a ToO event: if there are TAC-approved proposals for such events in the FOV and it is detected during an INTEGRAL observation, then the successful proposer will be granted the data rights on the event. Again, these are serendipitous events, and thus awarded zero time. If the positions of accepted targets are close-by, proposals could be merged into one observation with an optimised exposure time (called "amalgamation"), in order to save slewing time and to maximise the observation efficiency. Those proposals which were amalgamated to the main proposal were treated as obtaining zero time. In this chapter we only take into account Open Time proposals.

5.5   Time Allocation Committee (TAC)

The TAC reviews all submitted proposals and makes recommendations to the ESA Director of Science (D-SCI) on the observing programme to be performed by INTEGRAL. It takes into account the scientific case, justification, merit and relevance of the proposed observation(s), and the potential contribution of the overall scientific return of the mission. The TAC recommendations on the observing programme are given priority assignments of A, B or C, with A being the highest priority (in AO-1 and AO-2 accepted ToO proposals did not receive a grade and we treat them as being grade A). Priority A and B targets are of major scientific importance and are scheduled with the highest priority. Priority C targets are used as "fillers" and have a significant lower likelihood to be finally scheduled. Priority A and B targets are automatically transferred to the next observation period, if their successful observation commenced (i.e., at least 25% executed of the approved time), but is uncompleted during the current one. The total recommended observing time is about quarter more than the total available Open Time, for schedule planning efficiency. Approximately 45% of the recommended observing time is designated as Priority A, 30% as Priority B and 25% as Priority C. For ToO proposals, the actual requested observing time (other than the total time requested based on the number of targets) has been taken into account.

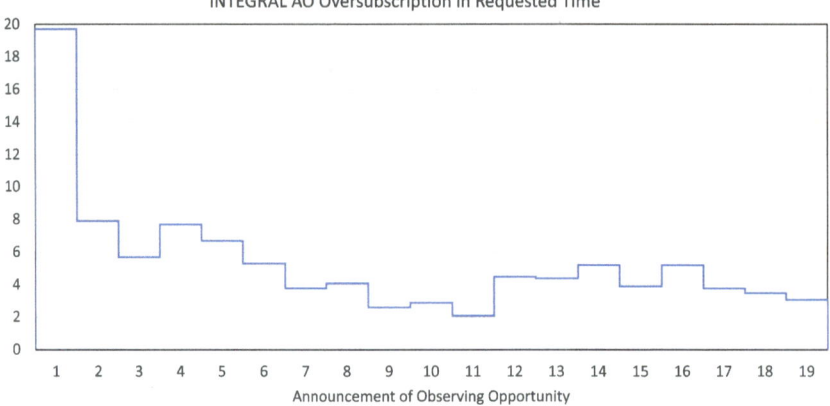

Fig. 5.1 The over-subscription of requested INTEGRAL observing time compared to that available for AO-1 to AO-19, covering an interval of 21 years from 2000

An average of 110 valid proposals were received in response to each AO (see Table 5.1). With the over-subscription[1] factors (see Fig. 5.1) scientific assessment of the proposals is a major undertaking of the astronomical community with over 340 scientists having participated so far. For the first two AOs, the TAC was divided into four panels to reflect the range of topics proposed. This was reduced to three panels for subsequent AOs. Much of the assessment is performed by panels, consisting typically of five scientists selected from the worldwide community. Each panel is led by a panel chair and the overall TAC is led by a chairperson. The names of the TAC members (with exception of the chairperson who sits on the INTEGRAL Users Group (IUG), see Sect. 5.6) are not made public.

The genders of the TAC members and panel chairs were assigned through the personal knowledge of the authors and SOC staff. We appreciate that gender identity is more complex than a binary issue. However, no attempt was made to assign genders other than male or female in this study. We then examined the gender composition of the TAC members and panel chairs (Table 5.1) from AO-1 to AO-19 covering 2001 to 2021. TAC Panel members are chosen for their high-level of relevant scientific knowledge, while selection of the TAC chairs was for scientists considered to have leadership roles in high-energy astronomy. Figure 5.2 shows the fraction of female INTEGRAL TAC members and Fig. 5.3 the same for the panel chairs. In total there were 226 male and 60 female panel members of which 46 males

---

[1] The over-subscription for INTEGRAL is defined as the ratio of the requested time over the available time. The available time is the total time expected to be available for Guest Observer observations in an upcoming AO observing period; for the later AOs this was about 21 Msec (note that this number does not include perigee passages, engineering observations, calibration observations, etc.). The requested time is the total time requested in non-ToO proposals plus 10% of the total time based on the number of targets asked in ToO proposals. The 10% is a kind of rough average probability that a ToO proposal is triggered.

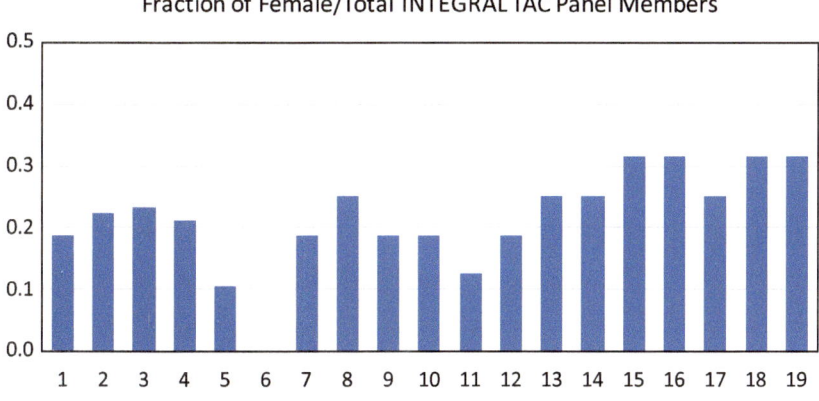

Fig. 5.2 The fraction of female INTEGRAL TAC members compared to the total for AO-1 to AO-19 covering an interval of around 21 years. An increasing trend in the fraction of females TAC members from ∼0.2 in AO-1 to ∼0.3 by AO-19 is evident

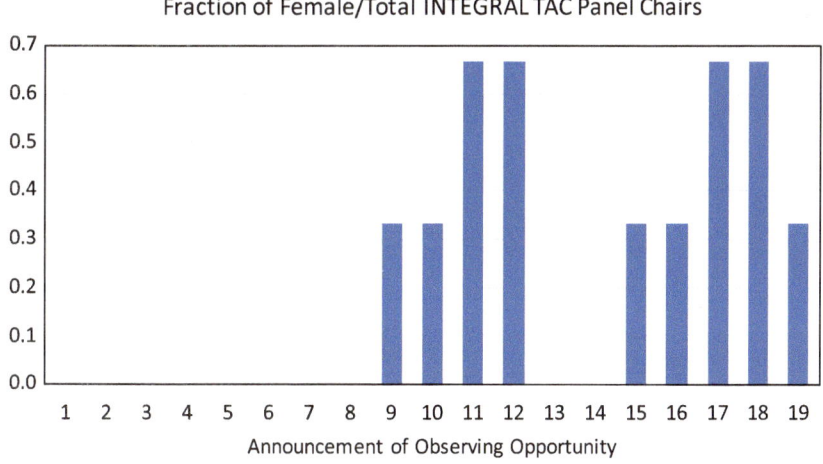

Fig. 5.3 The fraction of female INTEGRAL TAC chairs compared to the total for AO-1 to AO-19 covering an interval of around 21 years. We note that there were no female panel chairs until AO-9 in 2011

and 13 females were panel chairs. Since members and chairs may serve for multiple AOs, the numbers of individuals involved is smaller. Most of the TAC members (and chairs) are based at institutes located in the European Space Agency (ESA) Member

States, with a significant number from institutes in the Russian Federation[2] and the US. An increasing trend in the fraction of female TAC members from ~0.2 in AO-1 to ~0.4 by AO-19 is evident. Given the small number of panel chairs (59 in a total of 19 AOs), it is more difficult to quantify trends. However, it is worth noting that there were no female TAC panel chairs until AO-8 in 2010.

5.6   INTEGRAL Users Group

As well as the TAC, there is another body involved in optimising the scientific output of the mission. This is the INTEGRAL Users Group (IUG) which advises ESA, through the Project Scientist, on all matters relating to the optimisation of the scientific output of the mission. It also acts as a forum to discuss input from the community of users, and when appropriate advise or recommend action to ESA regarding INTEGRAL operations. The IUG is smaller than the XMM-Newton Users' Group (XUG) with only 6 members (including the chair). The names of the members are made public. The IUG was established in 2005, in parallel to the already existing ISWT. As a consequence of the termination of the CP beyond 2008, the IUG and ISWT teams were merged in November 2007 into one IUG. The chair of the INTEGRAL TAC is also a member of the IUG to ensure good coordination between the two bodies. In addition, the INTEGRAL Project Scientist, Mission Manager and instrument principal investigators attend the meetings. Note that there was a change in Project Scientist in 2013. Figure 5.4 shows the number of female IUG members divided by the total between 2005 and 2021. On average 28% of the total number of members were female with no obvious trend with time. We note that 53% of the IUG chairs were female.

5.7   Proposers

Between AO-1 and AO-19 there were 2115 proposals submitted in response to INTEGRAL calls for observing opportunities. This provides a rich data set to quantify many aspects of the proposal process. When submitting a proposal, INTEGRAL proposers are required to indicate the country where their institute or university is located. These are selected from a drop-down menu. This information allows an examination of the nationality distribution of the proposers' institutes. Proposers are not required to submit information on gender, age or type of position held and these have to be derived by other means if required.

---

[2] By agreement, 25% of total TAC members (per AO) must be from Russian Federation scientific community.

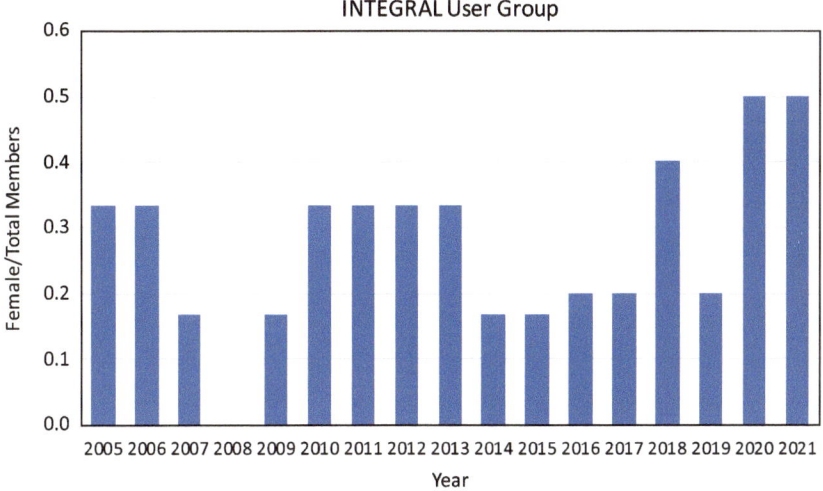

Fig. 5.4 The fraction of female INTEGRAL IUG members compared to the total membership between 2005 and 2021

5.7.1 Proposers' Institute Countries

Table 5.3 shows the number of proposals submitted and accepted for a range of countries including all the ESA Member States that submitted at least one proposal and other countries from which significant numbers of proposals originated. Only eight proposals were received from countries that are not listed in Table 5.3.

Scientists located at institutes and universities within Italy submitted the most proposals—nearly a quarter of the total (22.3%). This is followed by Germany (15.5%), the USA (12.8%), France (11.6%), the Russian Federation (8.9%), and Switzerland (7.0%). Figure 5.5 shows the percentages of proposal numbers versus AO for the six countries whose scientists submitted the most proposals. These show relatively stable proposal submission fractions with AO number. There is evidence for a decline in the percentage of proposals from Principal Investigator (PI)s located in the USA as the mission progressed (grey line in Fig. 5.5). This may indicate a lessening of interest in the mission from US based PIs as the mission progressed. We note that NASA funding for US INTEGRAL proposals was only available from AO-1 to AO-7.

In order to be able to draw reliable conclusions, we examined the number of accepted proposals for the seven countries whose scientists have submitted at least 100 proposals (see Table 5.3). The countries with the highest accepted fractions are Spain where 62.0% of 142 proposals were accepted, followed by the Russian Federation with 53.4% accepted of 189 proposals, and Germany with 47.3% accepted of 328 proposals.

Table 5.3 The number of proposals submitted and accepted from scientists located in different countries

Country	No. proposals submitted	Percentage of total	No. accepted	Acceptance percentage
BELGIUM	8	0.38	2	25.0
CZECH REPUBLIC	3	0.14	0	0.0
DENMARK	35	1.65	4	11.4
FINLAND	5	0.24	4	80.0
FRANCE	245	11.58	94	38.4
GERMANY	328	15.51	155	47.3
GREECE	1	0.05	0	0.0
IRELAND	28	1.32	1	3.6
ITALY	472	22.32	169	35.8
NETHERLANDS	70	3.31	33	47.1
POLAND	18	0.85	6	33.3
SPAIN	142	6.71	88	62.0
SWEDEN	4	0.19	0	0.0
SWITZERLAND	149	7.04	45	30.2
UNITED KINGDOM	74	3.50	33	44.6
AUSTRALIA	9	0.43	6	66.7
CHINA	21	0.99	10	47.6
INDIA	2	0.09	0	0.0
ISRAEL	1	0.05	0	0.0
JAPAN	18	0.85	4	22.2
MEXICO	1	0.05	0	0.0
RUSSIAN FEDERATION	189	8.94	101	53.4
TURKEY	14	0.66	9	64.3
UNITED STATES	270	12.77	107	39.6

5.7.2   Proposers' Gender

We next examined the number of proposals with male and female PIs from the same seven countries as discussed in Sect. 5.7.1 (Table 5.4). Since proposers are not asked to specify their gender (nor age or type of position held etc.) gender information for each proposer was obtained through examining publicly-accessible web-based data in a similar way as for the HST study by I.N. Reid [59] and from knowledge of the community they serve by the Project Scientist and SOC staff. We appreciate that gender identity is more complex than a binary issue, however, no attempt can be made to assign genders other than male or female as this information is not readily available.

The country with the highest fraction of female PIs is Italy (54.4%) followed by Spain (45.1%). The countries with the lowest fraction of female PIs are the Russian

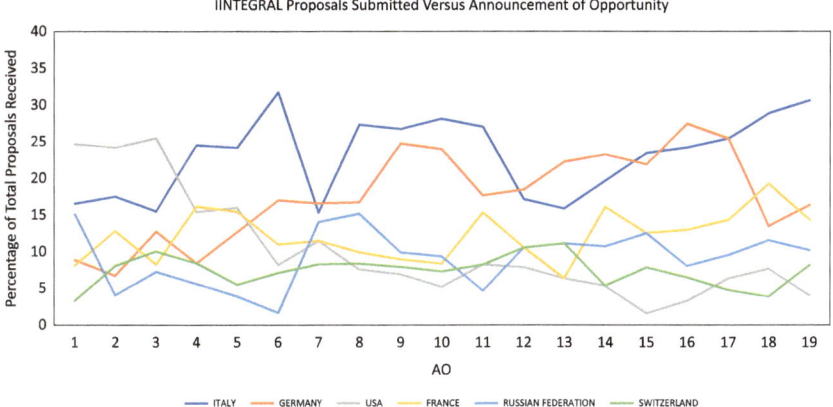

Fig. 5.5 Proposals submitted from the six countries with the largest numbers of proposals against AO number. There are no large changes in the percentages during the mission, suggesting that the level of interest from the six countries was unchanged. The only exception is a decline in the percentage of proposals from PIs located in the USA (grey line) as the mission progressed. Note that the Guaranteed Time ended in 2008 around the time of AO-6

Table 5.4 Proposal statistics for PIs located in countries with >100 proposals. The % differences are with respect to the acceptance percentages given in Table 5.3 which are for all proposals from each country

Country	Submitted		Female	Accepted		% Difference	
	Male PI	Fem. PI	/Total (%)	Male PI	Fem. PI	Male PI	Fem. PI
FRANCE	209	36	14.7	80	13	0.3	−1.8
GERMANY	281	47	14.3	139	16	2.2	−13.2
ITALY	215	257	54.4	92	77	7.4	−5.8
SPAIN	78	64	45.1	55	33	8.5	−10.4
SWITZERLAND	132	17	11.4	39	6	−0.7	5.1
RUSSIAN FEDERATION	189	0	0.0	101	0	0.0	...
USA	245	25	9.3	98	8	0.4	−7.6

Federation (0.0%) and the USA (9.3%). Figure 5.6 shows the acceptance fraction differences for male and female PIs compared to the average for that country for each of the seven countries with >100 proposals. It can be seen that Switzerland is the only country to have a better female PI acceptance rates of +5.1% compared to the average for all Swiss proposals. Female PIs located in Germany and Spain have acceptance rates of −13.2% and −10.4%, respectively compared to the averages for these countries. The accepted fraction difference for the Russian Federation comes out to 0, because the number of female PIs has been 0; see Table 5.4.

In Table 5.5 we collect per AO the number of proposals submitted, also per gender, as well as for the accepted proposals. Figure 5.7 shows that the fraction

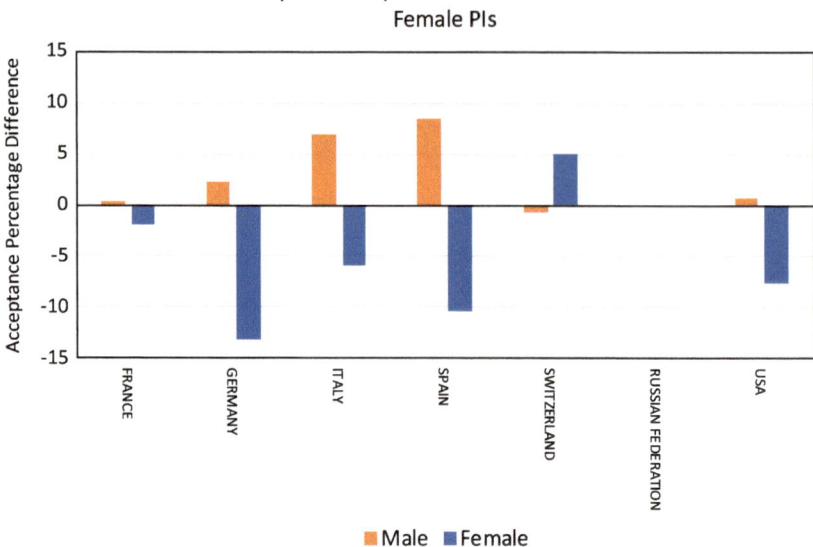

Fig. 5.6 The acceptance fraction differences of INTEGRAL proposals with male and female PIs compared to the average for all proposers from that country. The countries are those with at least 100 submitted INTEGRAL proposals

of proposals with female PIs compared to the total increased from about ~10% to about ~30% between AO-1 and AO-19.

5.8   Proposal Selection

We decided to only investigate the outcomes of proposals that requested observing time—so data rights or similar proposals were not included in the analysis of success rates. Data rights proposals have different factors affecting their selection and so their success rates cannot be directly compared with other proposals. In addition, data right proposals are a peculiarity of the INTEGRAL mission and cannot easily be compared with other missions. We note that, examination of the 607 data rights proposals shows that 511 were awarded data rights. This is 84% of the proposals, a much higher percentage than for proposals requesting dedicated observing time. For the proposals requesting observing time, we investigated the number of accepted proposals and the amount of observing time allocated. We examined the statistics for all proposals—Priority A, B and C and only for those ranked in Priority A and B by the TAC since these are of major scientific importance and had a much higher chance of being scheduled than the "filler" Priority C targets.

Table 5.5 INTEGRAL AO proposal numbers showing for proposals with male and female PIs and the number of proposals accepted. Only proposals that requested observing time are included

| | Submitted | | | Accepted | |
| | No. of | Male PI | Female PI | Male PI | Female PI |
AO	proposals	proposals	proposals	proposals	proposals
1	265	237	28	75	15
2	143	124	19	71	12
3	103	85	18	48	11
4	93	82	11	40	5
5	93	79	14	49	5
6	70	59	11	35	8
7	70	59	11	38	5
8	61	51	10	32	6
9	46	39	7	33	5
10	57	46	11	31	7
11	53	40	13	32	10
12	72	54	18	29	8
13	57	39	18	23	9
14	54	34	20	27	11
15	61	43	18	31	9
16	58	39	19	27	7
17	59	37	22	29	14
18	48	32	16	27	10
19	45	31	14	27	9
Total	**1508**	**1210**	**298**	**704**	**166**

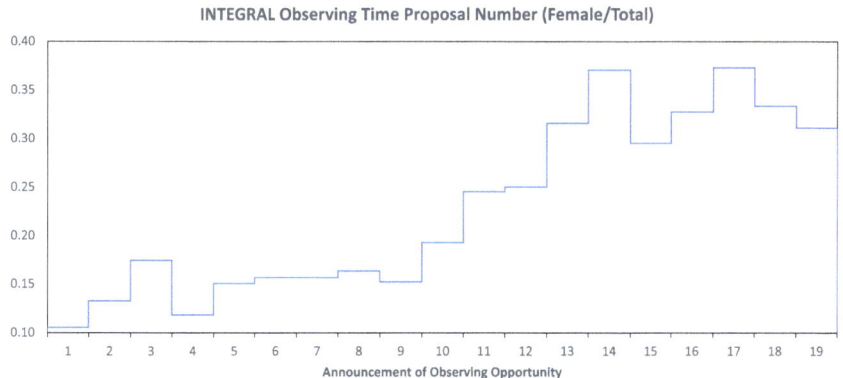

Fig. 5.7 The fraction of submitted INTEGRAL proposals with female PIs compared to the total. This increased from ~10% to ~30% between AO-1 and AO-19

5.8.1 Proposers' Age

In order to determine the "academic age" of proposers we used the difference between the year that their Doctor of Philosophy Degree (PhD) was awarded and the year that an AO to which they submitted a proposal was issued. Thus a proposer who submits multiple proposals to the same AO will be counted multiple times in the same "academic year", whilst one who submits multiple proposals to different AOs will be counted in different "academic years".

The year the proposer obtained their PhD (or equivalent) were determined for 99.7% of the proposals by searching the internet, particularly sites such as the Astrophysics Data Service (ADS), LinkedIn, the Astronomy Genealogy Project (astrogen.aas.org), ORCID.org, IEEE Xplore (https://ieeexplore.ieee.org/Xplore/home.jsp), and, for French theses https://www.theses.fr. Some proposers were contacted and provided their PhD dates by email. The proposers for which the PhD could not be found are often retired, deceased or have left astronomy. For PhD students who had not yet completed their degrees, the expected year of submission was used. For the small number of proposers who did not have a PhD and were not enrolled in a PhD programme, their dates were assumed to be arbitrarily far in the future. For some senior scientists in Italian institutes who do not have a PhD, three years after they obtained their Laurea was used. It should be noted that using the year of PhD to indicate the number of years experience neglects time spent outside of astronomy. We note that the "youngest" proposer was 13 years before obtaining a PhD and the oldest 53 years after.

5.8.2 Priority A, B and C Proposal Acceptance Rates

Of the 1508 proposals which requested observing time that were submitted to AO-1 to AO-19, 870 were awarded an amount of observing time. This corresponds to an overall success rate of 57.7%. This is higher than for XMM-Newton (Chap. 4), consistent with the more moderate over-subscription factors given in Table 5.1 compared to Table 4.1. We next examined the number of proposals for each gender that were awarded any observing time. A total of 1210 such proposals with male PIs and 298 proposals with female PIs were submitted, of which 704 and 166 were awarded observing time (see Table 5.5). This gives success rates of 58.2 and 55.7% for male and female PIs, respectively. This is a difference in favour of males of 4.4%.

Proposal submissions and proposal acceptance are unlikely to be independent or random processes and for Poisson statistics to apply events need to be independent of each other. If this is assumed to be the case then the success rates are 58.2 ± 2.8 and $55.7 \pm 5.4\%$ for male and female PI proposals, respectively and a difference in favour of males of $4.4 \pm 6.7\%$ corresponding to a male to female success ratio of 1.053 ± 0.113. These uncertainties are given only to illustrate the outcomes if

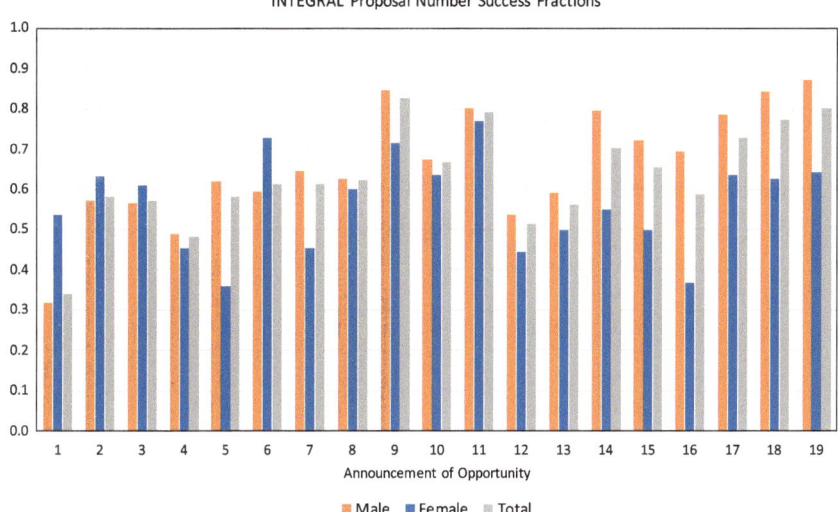

Fig. 5.8 The fraction of accepted proposals compared to the number submitted. The histograms show the acceptance fractions for male PIs, female PIs and all PIs

Poisson statistics were to apply to the selection process. Assuming this is correct, the success rates are not significantly different between the two genders. We note that actual uncertainties on the proposal numbers for each AO are zero. We further note that there are systematic uncertainties associated with this process due to misappropriated genders and incorrect PhD "academic ages", but these are likely to be too small to significantly affect our calculations.

The success rates are shown for male, female and all proposers for each AO in Fig. 5.8. In contrast to the results from HST Cycles 11 to 20 reported in [59] where proposals with male PIs had a consistently higher success rate than those with female PIs, no such trend is visible in INTEGRAL proposals. There is no obvious evolution in the male and female PI proposal acceptance rates with AO number, which cover an interval of more than 20 years.

Figure 5.9 shows the variation in acceptance rates more clearly. For each AO, it shows the difference between the expected number of accepted proposals, calculated using the overall acceptance rate, and the number actually accepted. It shows that the largest discrepancies occurred during AO-1 and AO-2 where proposals with female PIs had higher success rates of 17.9 and 6.6%, respectively, compared to proposals with a male PI. The most successful AO for male PI proposals was AO-19 where proposals led by male PIs were 4.9% more likely to be accepted than those with a female PI.

We examined the TAC membership genders to see if these are correlated with proposal acceptance rates. Figure 5.10 shows the acceptance fraction of proposals with female PIs against the fraction of female compared to the total number of TAC members. It can be seen that there is no obvious correlation between the success

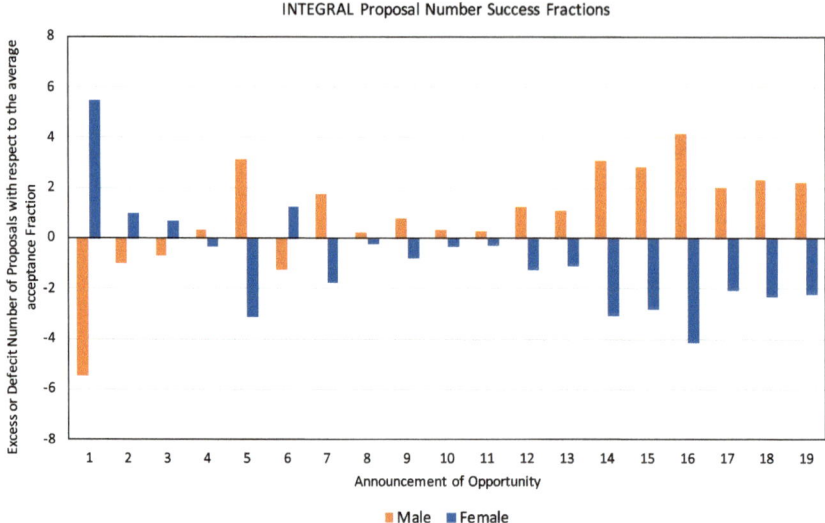

Fig. 5.9 The relative success rates of INTEGRAL proposals.The histograms show the difference between the actual number of successful proposals and the expected number based on the overall acceptance rate

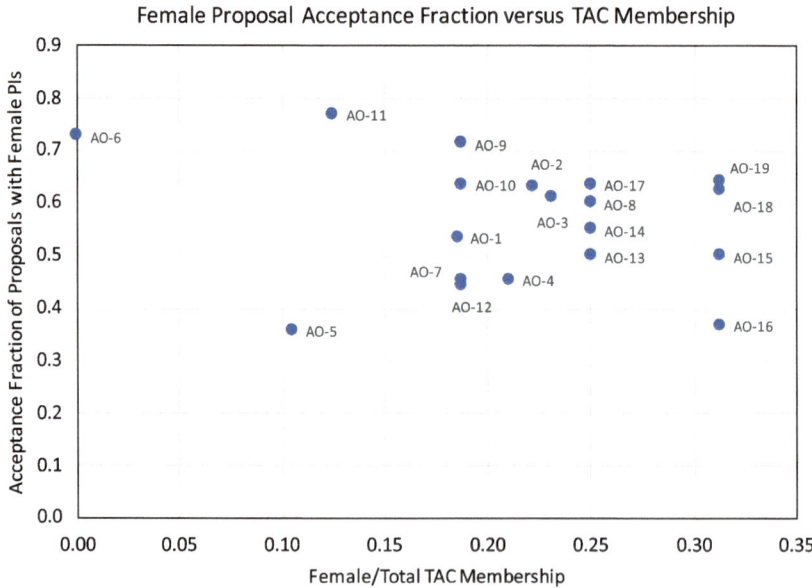

Fig. 5.10 The acceptance fraction for female PIs for each AO (labelled) against the fraction compared to the total of female TAC members. There is no obvious correlation between the two, or with the number of female panel chairs

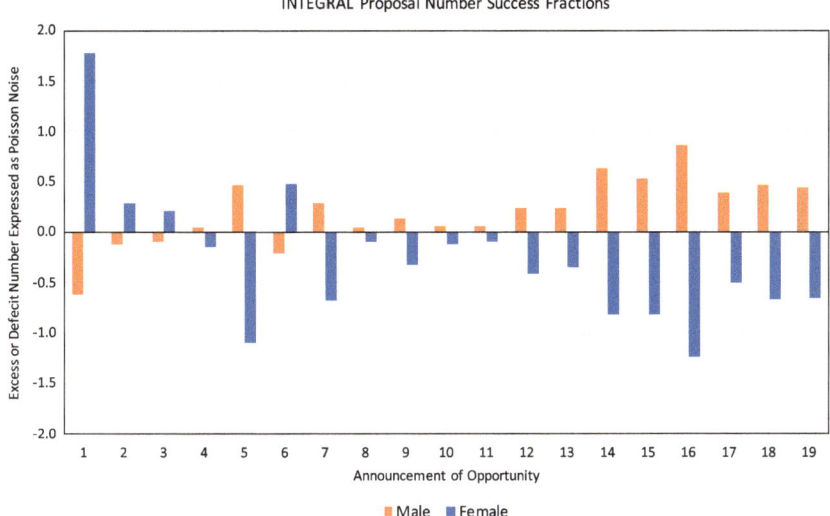

Fig. 5.11 The relative success rates of INTEGRAL proposals. The histograms show the difference between the actual number of successful proposals and the expected number based on the overall acceptance rate normalised by dividing by the square root of the number of expected proposals

rates of female-led proposals and the fraction of female TAC members. We note that there were no female chairs for AO-1 to AO-8, or AO-13 and AO-14 (see Fig. 5.3) and the results from these AOs are indistinguishable from the other AOs suggesting that having females in leadership positions in the TAC does not make a measurable difference to female PI proposal success rates.

Figure 5.11 shows the same excesses depicted as a quasi-Poisson quantity, normalised by dividing by the square root of the expected number of accepted proposals. While Poisson statistics do not necessarily apply to the peer-review process, scientists often adopt such a statistic in assessing the difference between observation and prediction. It is interesting to note that from the 19 AOs examined, 9 had higher female PI success rates than for male PIs, while the opposite occurred during the other 10 AOs.

5.8.3   Priority A and B Proposal Acceptance Rates

Of the 1508 proposals which requested observing time that were submitted to AO-1 to AO-19, 690 were awarded Priority A or B observing time. This corresponds to an overall success rate of 45.8%. We next examined the number of proposals for each gender that were awarded high-priority observing time. A total of 1210 such proposals with male PIs and 298 proposals with female PIs were submitted, of which 556 and 134 proposals were awarded observing time. This gives success rates of 46.0 and 45.0% for male and female PIs, respectively. This is a difference in favour of males of 2.2%. Again, assuming square root errors on the numbers of submitted and accepted proposals gives success rates of 46.0±2.4% and 45.0±4.7% for male and female PI proposals, respectively. This corresponds to a male to female success ratio of 1.022 ± 0.118. Again, given the assumed errors, the success rates are not significantly different between the two genders.

5.8.4   Priority A, B and C Observing Time Acceptance Rates

We examined the acceptance rates for the genders using awarded times. Many INTEGRAL proposals are allocated less observing time than requested, often because not all the requested targets are approved. If the allocation of observing time is handled differently for males and female PIs, then this will show as a difference in the relative amounts of time approved, compared to the number of proposals accepted. The results for each AO are summarised in Table 5.6. This

Table 5.6 INTEGRAL requested and accepted observing time for male and female proposers between AO-1 and AO-19. The last two columns give the ratio of accepted to proposed observing time for each gender. A proposal with a male PI which requested 1000 Msec of observing time in AO-1 has been excluded

| | Time (Msec) | | | | Success Rate | |
| | Requested | | Accepted | | Time | |
AO	Male	Female	Male	Female	Male	Female
1	283.0	39.8	32.7	5.4	0.116	0.135
2	132.4	14.1	49.4	5.7	0.373	0.402
3	102.3	12.7	48.3	6.7	0.472	0.526
4	121.0	15.5	27.6	6.3	0.228	0.406
5	132.9	23.6	46.0	6.4	0.346	0.272
6	75.0	11.6	26.8	5.1	0.357	0.439
7	92.5	11.3	41.0	5.7	0.443	0.499
8	71.9	14.1	33.1	7.4	0.461	0.527
9	52.3	10.1	32.9	7.1	0.628	0.701
10	48.7	9.8	30.2	6.8	0.619	0.698
11	36.1	14.5	29.0	12.2	0.805	0.838
12	77.6	21.6	33.3	7.4	0.430	0.342
13	81.0	21.7	27.7	9.0	0.342	0.415
14	74.3	37.2	25.6	9.6	0.345	0.258
15	47.9	33.5	34.8	8.6	0.726	0.258
16	76.3	33.2	32.1	7.9	0.421	0.237
17	44.5	36.0	32.7	12.0	0.735	0.332
18	38.2	36.2	35.3	10.5	0.923	0.290
19	38.3	26.8	37.2	13.2	0.971	0.492
Total	1626.2	423.2	655.6	152.7		

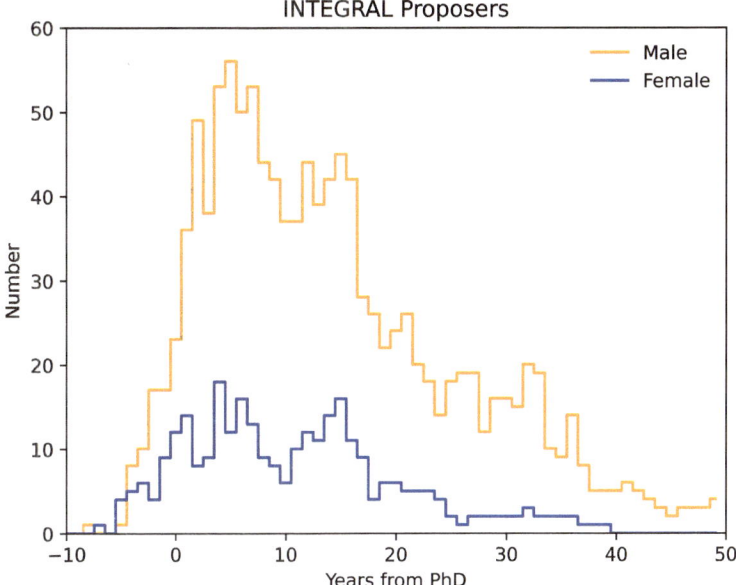

Fig. 5.12 The "academic age" distribution (Years since PhD) for male and female INTEGRAL PIs. PIs have a mean "academic age" of 13.7 years. Male PIs have a mean "academic age" of 14.7 years compared to 10.8 years for female PIs

shows the total time requested by proposals with male and female PIs to be 2626.2 and 423.2 Msec, respectively. 655.6 Msec and 152.7 Msec of observing time was awarded to proposals with male and female PIs, respectively. This gives average success rate of 40.3% for male PIs and 36.1% for female PIs. This is a difference of 11.5% in favour of male PIs. We note that a single proposal with a male PI which requested 1000 Msec of observing time (or about 32 years!) and was awarded 1 Msec (in Priority C in AO-1) has been excluded from this analysis as the request was deemed unrealistic.

Figure 5.12 shows the distributions of male and female INTEGRAL PIs who obtained their PhDs or equivalent in one year bins between 10 years before the PhD date to 50 years after. The male PIs who requested observing time (i.e., data time proposals are excluded) have a mean "academic age" of 14.7 years post-PhD compared to that of the female PIs of 10.8 years. The mean for all PIs is 13.7 years. A two-sample Anderson-Darling Test (e.g., [60]) shows that the hypothesis that both samples come from the same underlying population can be rejected at >99% confidence. This difference can be more clearly seen in Fig. 5.13 which shows the ratio of the two curves. There appears to be an increase in the number of early career (less than around 5 years after obtaining a PhD) female PIs compared to males. Beyond this the ratio of females to the total is ∼20%.

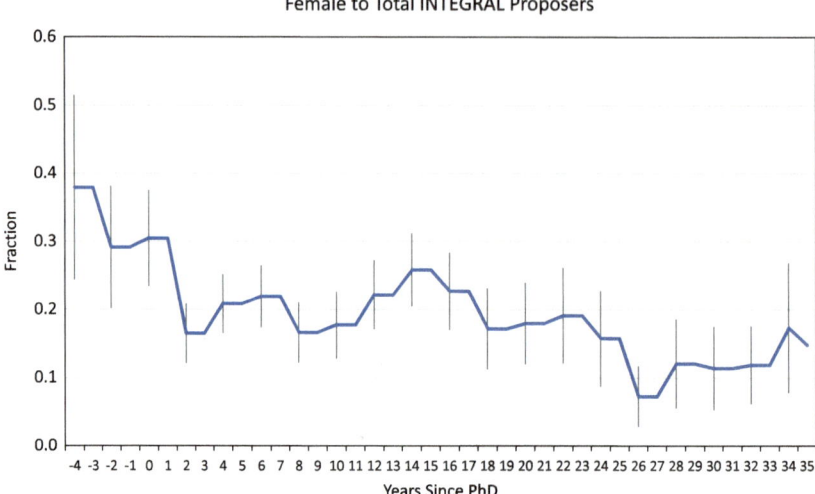

Fig. 5.13 The ratio of the "academic age" (years from PhD) of female to all INTEGRAL PIs in two year bins showing an increase in early career (less than around 5 years after obtaining a PhD) female PIs compared to the total. The indicative error bars show 1σ standard deviations assuming that the number of proposers in each age bin follows a Poisson distribution

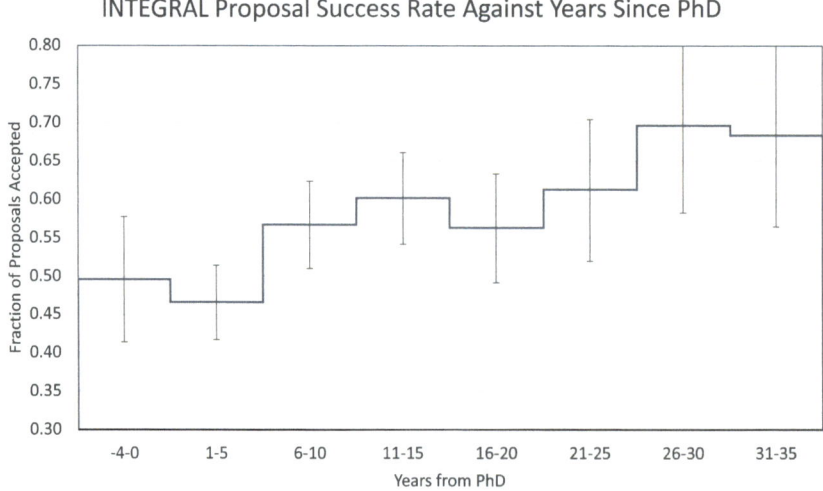

Fig. 5.14 The acceptance fraction of all INTEGRAL PIs with year of their PhD. This shows an increase in the acceptance rate from $\sim$45% for PhD students to $\sim$70% for "senior" researchers. The indicative error bars show 1σ standard deviations assuming that the number of proposers in each age bin follows a Poisson distribution

We next investigated the acceptance rates for the proposals against PhD year. This is shown for years −4 to 35 years after PhD in Fig. 5.14. A least squares fit to the data gives an intercept of $(45.7 \pm 0.4)\%$ and a gradient of $(0.73 \pm 0.26)\%$ year^{-1} assuming that the uncertainties are the square root of the number of people in each bin. The value of R^2 is 0.45. Indicative error bars show 1σ standard deviations assuming that the number of proposers in each age bin follows a Poisson distribution. There as an increase in the acceptance rate from $\sim$45% for PhD students to $\sim$70% for late career researchers. We note that this gradient is almost a factor two stronger than measured with XMM-Newton. This increase could result from a combination of factors including:

1. An increase in the success rate of researchers as they gain experience and have expanded networks of collaborators.
2. Less successful proposers who leave astronomy.
3. Less successful proposers who remain in astronomy, but do not propose in subsequent INTEGRAL AOs.
4. A bias in the selection process towards late career scientists.

The positive gradient of this relation indicates that the female early career proposer population is less likely to have proposals accepted simply due to having less experience. To calculate the size of this effect, we took the male and female PI distributions shown in Fig. 5.12 and multiplied the number of PIs each year by the linear value derived from the fit to the overall success rates. This gave a predicted difference in the acceptance ratio of 2.4% for male and female PIs. This indicates that the underlying "academic age" differences between the two populations may be responsible for about half of the difference in gender acceptance rates of about 4.4% in proposal numbers or a quarter of the difference in accepted time (11.5%).

The acceptance fractions of female and male INTEGRAL PIs with the binned year of their PhD are shown in Fig. 5.15. The proposal success rate of female PIs tends to increase with "academic age" whilst that of male PIs shows a smaller increase, or may be constant. This is in contrast to results reported from HST [59] and XMM-Newton (Chap. 4) where the success rates for late career ($\sim$20 years post PhD) female PIs appear to be lower than their early career counterparts. Indicative error bars show 1σ standard deviations assuming that the number of proposers in each age bin follows a Poisson distribution. Error bars are only shown for female PIs as these are much larger than for male PIs due to the smaller number of female PIs.

5.8.5 Priority A and B Observing Time Acceptance Rates

We next examined the acceptance rates for the genders awarded high-priority (A or B) observing time. The total time requested by proposals with male and female PIs is 2626.2 and 423.2 Msec, respectively. 551.1 Msec and 139.2 Msec of Priority A or B observing time was awarded to proposals with male and female PIs, respectively.

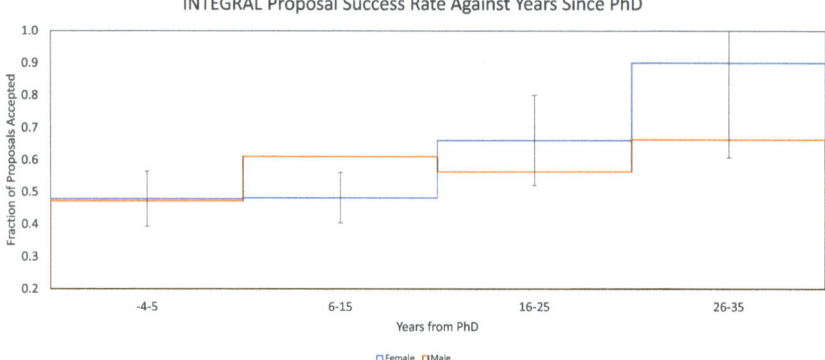

Fig. 5.15 The acceptance fractions of female and male INTEGRAL PIs with year of their PhD. The proposal success rate of females PIs tends to increase with "academic age" whilst that of male PIs shows a smaller increase, or may be constant. The indicative error bars show 1σ standard deviations assuming that the number of proposers in each age bin follows a Poisson distribution. Error bars are only shown for female PIs as these are much larger than for male PIs due to the smaller number of female PIs

This gives average success rate of 33.9% for male PIs and 32.9% for female PIs. This is a difference of 2.9% in favour of male PIs. As before, a single proposal with a male PI which requested 1000 Msec of observing time (or about 32 years!) and was awarded 1 Msec (in Priority C in AO-1) has been excluded from this analysis.

5.9   Proposal Selection Summary

We have examined the outcomes of 1508 proposals submitted to ESA requesting observing time in response to the AO-1 to AO-19 calls for INTEGRAL observing time. The proposal acceptance rates are summarised in Table 5.7. A total of 58.2% of proposals with male PIs and 55.7% with female PIs were awarded *any* observing time on INTEGRAL. This is a difference in favour of males of 4.4%. When the amount of awarded time is evaluated this changes to being 11.5% in favour of proposals with male PIs. This difference implies that proposals with female PIs were awarded a smaller fraction of the requested observing time than their male counterparts. For the proposals that were awarded Priority A or B time, the difference in the number of accepted proposals is 2.2% and in terms of awarded time 2.9% in favour of male PIs. Both of these differences are smaller than when Priority C time is included. This implies that female PIs have much lower success rates than their male counterparts for proposals awarded Priority C time. We interpret these four different numbers to imply an overall male to female success rate difference of 2–11% for INTEGRAL proposals.

The acceptance rate for data rights proposals was 84%—much higher than for proposals requesting observing time. This suggests that submitting data rights proposals is a good strategy for obtaining INTEGRAL data.

Table 5.7 Summary of INTEGRAL Proposal Acceptance Rates for AO-1 to AO-19 for proposals requesting observing time. A proposal with a male PI which requested 1000 Mec of observing time in AO-1 has been excluded from the analysis. The uncertainties given in brackets assume that the number of submitted and accepted proposals have square root uncertainties and for guidance only

		Proposal PI		
Priority	Parameter	All	Male	Female
	Proposals submitted	1508	1208	300
A, B, C	Number accepted	870	704	166
	Percentage accepted	57.7	58.2	55.7
	Ratio male/female		1.044	
A, B	Number accepted	690	556	134
	Percentage accepted	45.8	46.0	45.0
	Ratio male/female		1.022	
	Time requested (Msec)	2049.4	1626.2	422.7
A, B, C	Time accepted (Msec)	808.3	655.6	152.7
	Percentage time accepted	39.4	40.3	36.1
	Ratio male/female		1.115	
A, B	Time accepted (Msec)	690.2	551.1	139.2
	Percentage time accepted	33.7	33.9	32.9
	Ratio male/female		1.029	
	Overall male/female difference		2–11%	

Scientists located at institutes within Italy submitted the most proposals—almost a quarter of the total followed by those located in institutes in Germany, the USA and Switzerland. Amongst the seven countries submitting >100 proposals, Spain had 62% of 142 proposals accepted, followed by the Russian Federation with 53% of 189 proposals, and Germany with 47% of 328 proposals. Italy is the country with the highest fraction of proposals from female PIs (55%) followed by Spain (45%). The countries with the lowest fraction of proposals from female PIs are the Russian Federation (0%) and the USA (9%). Switzerland is the only country to have a better female PI proposal acceptance rate than for male PIs of +5% compared to the average for all Swiss proposals. Proposals from female PIs located in Germany and Spain have acceptance rates of −13% and −10%, respectively compared to the averages for their countries.

The fraction of proposals from female PIs increased from ~10% of the total during the early AOs to ~30% of the total in the later AOs. However, there is no marked evolution in the gender balance of the proposal acceptance rates with AO number and hence between 2000 and 2021. There was an increase in the fraction of female TAC members from ~0.2 of the total for the first AOs to ~0.3 for the latest AOs and a larger increase in the number of female TAC panel chairs from none before AO-9 to one or two out of three chairs in subsequent AOs. These results suggest that increasing fractions of female TAC members or panel chairs does not make a measurable difference to the success rate of female PI proposals.

Using the year of obtaining a PhD, or equivalent degree, as a proxy for "academic age" shows that the mean "academic age" of female PIs is 3.9 years less than for males. A linear fit to the proposal acceptance rate against PhD year shows an increase in acceptance rate from ∼45% for students to ∼70% for senior astronomers. The proposal success rate of females PIs tends to increase with "academic age" whilst that of male PIs shows a smaller increase, or may be constant. This is in contrast to results reported from HST [59] and XMM-Newton (Chap. 4) where the success rates for late career (∼20 years post PhD) female PIs appear to be lower than their early career counterparts.

5.10   Discussion

It is interesting to compare our results with those from other missions and facilities. XMM-Newton and HST are the obvious missions for comparison as they are similarly long-lived as INTEGRAL and both have much higher average numbers of proposals in response to each AO. There have also been investigations of the outcomes of the observing selection processes for European Southern Observatory (ESO), Canadian facilities and the National Radio Astronomy Observatory (NRAO) facilities, which are also discussed in Sect. 4.9.

For XMM-Newton the difference observed in the proposal selection process of 5–15% in favour of male PIs is comparable to that observed for INTEGRAL of 2–11%. Both of these ranges are small in comparison to those reported for some ESO, and NRAO facilities [61–63] and for HST prior to the implementation of dual anonymous reviewing. [59] reports on a study of gender-based systematic trends in the HST proposal review process for Cycles 11 through 21 (2001 to 2013). From this study of the outcomes of 9400 proposals there appears to be a similar trend to that seen here in that male PIs are more likely to succeed in achieving a successful HST proposal (23.5% success rate) than female PIs (18.1% success rate). This is a difference of 30% in favour of proposals led by male PIs. With dual anonymous reviewing the names of the reviewers and the investigators are made known to each other only after the review has been completed. Once dual-anonymous reviewing was implemented for HST the male and female PI proposal acceptance difference falls to 9.3% [64], comparable to that seen with INTEGRAL and XMM-Newton.

An investigation of the time allocation process at ESO is reported in [61]. This covers an interval of 8 years and involved about 3000 PIs. Female PIs were found to have a significantly lower chance of being awarded a top rank compared to male PIs with a male/female ratio of 1.39 ± 0.05. The paper suggests that the principal explanation for the difference may be due to the average higher seniority of the male PIs assuming that more senior scientists of both genders write better proposals and thus succeed more often at obtaining ESO telescope time. However, no attempt was made to quantify if this effect can account for the observed differences. Nevertheless, [61] concludes that the ESO review process itself introduces extra gender differences.

An investigation of the gender-related systematics in the proposal review processes for the four facilities operated by NRAO: the Jansky Very Large Array (JVLA), the Very Long Baseline Array (VLBA), the Green Bank Telescope (GBT) and Atacama Large Millimetre/sub-millimetre Array (ALMA) are reported in [62] and [63]. Similarly to HST, before the introduction of double-anonymous reviewing, and similarly as well to the ESO study there are significant gender-related effects in the proposal rankings in favour of male PIs compared to female PIs for all four facilities with varying degrees of confidence reflecting the different number of proposals.

In summary, the difference observed in the INTEGRAL proposal selection process of 2–11% in favour of male PIs is smaller than those reported for some ESO, and NRAO facilities [61, 62] and for HST prior to the implementation of dual anonymous reviewing [59]. Once dual-anonymous reviewing was implemented for HST the male and female PI proposal acceptance difference falls to 9.3% [64], comparable to that seen here with INTEGRAL of 2–11%. This suggests that the INTEGRAL proposal selection processes produces outcomes consistent with, or at least close to, the best available science. It does not, however, provide gender parity and it would be useful to investigate reasons for this. The acceptance rates of male and female proposers for different geographical regions could be examined. This could reveal that regional differences play a role, perhaps related to language skills. It would also be useful to investigate the uncertainties in the proposal selection process. This could be accomplished by comparing the results of different representative TACs evaluating the same set of proposals. Finally, it would be interesting to examine the publication rates of early and late career scientists who have been awarded INTEGRAL observing time. Are late career scientists better at exploiting their observations and so produce relatively more publications and citations than their early career colleagues? More generally, what is the effect on the science return of a mission if the observations are dominated by late or early career scientists? This is a complex issue as, e.g., a late career scientist may have made substantial contributions to a proposal from an early career colleague, so improving its chances of success.

Acknowledgments We thank the INTEGRAL SOC and ISDC staff for their support. This research was supported by the International Space Science Institute (ISSI) in Bern, through the ISSI Working Group project 'The Scientific Performance of ESA's Science Missions'.

References

1. C. Winkler, T.J.L. Courvoisier, G. Di Cocco, et al., The INTEGRAL mission. Astron. Astrophys. **411**, L1–L6 (2003)
2. E. Kuulkers, C. Ferrigno, P. Kretschmar, et al., INTEGRAL reloaded: spacecraft, instruments and ground system. New Astron. Rev. **93**, 101629 (2021)
3. P. Ubertini, F. Lebrun, G. Di Cocco, et al., IBIS: the imager on-board INTEGRAL. Astron. Astrophys. **411**, L131–L139 (2003)

4. G. Vedrenne, J.P. Roques, V. Schönfelder, et al., SPI: the spectrometer aboard INTEGRAL. Astron. Astrophys. **411**, L63–L70
5. N. Lund, C. Budtz-Jørgensen, N.J. Westergaard, et al., JEM-X: the X-ray monitor aboard INTEGRAL. Astron. Astrophys. **411**, L231–L238 (2003)
6. J.M. Mas-Hesse, A. Giménez, J.L. Culhane, et al., OMC: an optical monitoring camera for INTEGRAL. Instrument description and performance. Astron. Astrophys. **411**, L261–L268 (2003)
7. T.J.L. Courvoisier, R. Walter, V. Beckmann, et al., The INTEGRAL science data centre (ISDC). Astron. Astrophys. **411**, L53–L57 (2003)
8. E.P.J. van den Heuvel, Gamma-ray observatory INTEGRAL reloaded. Nat. Astron. **1**, 0083 (2017)
9. S. Mereghetti, V. Savchenko, C. Ferrigno, et al., INTEGRAL discovery of a burst with associated radio emission from the magnetar SGR 1935+2154. Astrophys. J. Lett. **898**(2), L29 (2020)
10. C. Ferrigno, V. Savchenko, A. Coleiro, et al., Multi-messenger astronomy with INTEGRAL. New Astron. Rev. **92**, 101595 (2021)
11. A.J. Bird, A. Bazzano, A. Malizia, et al., The IBIS soft gamma-ray sky after 1000 integral orbits. Astrophys. J. Suppl. **223**(1), 15 (2016)
12. R.A. Krivonos, S.Y. Sazonov, E.A. Kuznetsova, et al., INTEGRAL/IBIS 17-yr hard X-ray all-sky survey. Month. Not. R. Astron. Soc. **510**(4), 4796–4807 (2022)
13. T. Siegert, R. Diehl, G. Khachatryan, et al., Gamma-ray spectroscopy of positron annihilation in the Milky Way. Astron. Astrophys. **586**, A84 (2016)
14. W. Wang, T. Siegert, Z.G. Dai, et al., Gamma-Ray emission of 60Fe and 26Al radioactivity in our galaxy. Astrophys. J. **889**(2), 169 (2020)
15. T. Siegert, R.M. Crocker, O. Macias, et al., Measuring the smearing of the Galactic 511-keV signal: positron propagation or supernova kicks? Month. Not. R. Astron. Soc. **509**(1), L11–L16 (2022)
16. M. Ajello, W.B. Atwood, M. Axelsson, et al., Fermi large area telescope performance after 10 years of operation. Astrophys. J. Suppl. **256**(1), 12 (2021)
17. B.P. Abbott, R. Abbott, T.D. Abbott, et al., Gravitational waves and gamma-rays from a binary neutron star merger: GW170817 and GRB 170817A. Astrophys. J. Lett. **848**(2), L13 (2017)
18. V. Savchenko, C. Ferrigno, E. Kuulkers, et al., INTEGRAL detection of the first prompt gamma-ray signal coincident with the gravitational-wave event GW170817. Astrophys. J. Lett. **848**(2), L15 (2017)
19. J. Rodriguez, M. Cadolle Bel, J. Alfonso-Garzón, et al., Correlated optical, X-ray, and γ-ray flaring activity seen with INTEGRAL during the 2015 outburst of V404 Cygni. Astron. Astrophys. **581**, L9 (2015)
20. J.P. Roques, E. Jourdain, A. Bazzano, et al., First INTEGRAL observations of V404 cygni during the 2015 outburst: spectral behavior in the 20-650 keV energy range. Astrophys. J. Lett. **813**(1), L22 (2015)
21. L. Natalucci, M. Fiocchi, A. Bazzano, et al., High energy spectral evolution of V404 cygni during the 2015 June outburst as observed by INTEGRAL. Astrophys. J. Lett. **813**(1), L21 (2015)
22. E. Jourdain, J.P. Roques, J. Rodi, A challenging view of the 2015 summer V404 Cyg outburst at high energy with INTEGRAL/SPI: the finale. Astrophys. J. **834**(2), 130 (2017)
23. J. Rodi, E. Jourdain, J.P. Roques, Timing analysis of V404 Cyg during its brightest outburst with INTEGRAL/SPI. Astrophys. J. **848**(1), 3 (2017)
24. C. Sánchez-Fernández, J.J.E. Kajava, S.E. Motta, et al., Hard X-ray variability of V404 Cygni during the 2015 outburst. Astron. Astrophys. **602**, A40 (2017)
25. J.J.E. Kajava, S.E. Motta, C. Sánchez-Fernández, et al., The December 2015 rebrightening of V404 Cygni. Variable absorption from the accretion disc outflow. Astron. Astrophys. **616**, A129 (2018)
26. J.P. Roques, E. Jourdain, On the high-energy emissions of compact objects observed with INTEGRAL SPI: event selection impact on source spectra and scientific results for the bright

sources Crab Nebula, GS 2023+338 and MAXI J1820+070. Astrophys. J. **870**(2), 92 (2019)

27. J.J.E. Kajava, C. Sánchez-Fernández, J. Alfonso-Garzón, et al., Rapid spectral transition of the black hole binary V404 Cygni. Astron. Astrophys. **634**, A94 (2020)

28. F. Cangemi, J. Rodriguez, V. Grinberg, et al., INTEGRAL discovery of a high-energy tail in the microquasar Cygnus X-3. Astron. Astrophys. **645**, A60 (2021)

29. J. Rodi, A. Tramacere, F. Onori, et al., A broadband view on microquasar MAXI J1820+070 during the 2018 outburst. Astrophys. J. **910**(1), 21 (2021)

30. M. Gilfanov, R. Syunyaev, E. Churazov, et al., Observations of Nova MUSCAE with the sigma telescope on the GRANAT observatory - spectroscopic properties in hard X-Rays and discovery of the annihilation line in the spectrum. Soviet Astron. Lett. **17**, 437 (1991)

31. A. Goldwurm, J. Ballet, B. Cordier, et al., SIGMA/GRANAT soft gamma-ray observations of the X-Ray nova in musca: discovery of positron annihilation emission line. Astrophys. J. Lett. **389** L79 (1992)

32. T. Siegert, R. Diehl, J. Greiner, et al., Positron annihilation signatures associated with the outburst of the microquasar V404 Cygni. Nature **531**(7594), 341–343 (2016)

33. P. Laurent, C. Gouiffes, J. Rodriguez, et al., INTEGRAL/IBIS observations of V404 Cygni polarimetric properties during its 2015 giant flares. PoS(INTEGRAL2016) **1**(285), 022 (2017)

34. E. Churazov, R. Sunyaev, J. Isern, et al., Cobalt-56 γ-ray emission lines from the type Ia supernova 2014J. Nature **512**(7515), 406–408 (2014)

35. R. Diehl, T. Siegert, W. Hillebrandt, et al., Early 56Ni decay gamma rays from SN2014J suggest an unusual explosion. Science **345**(6201), 1162–1165 (2014)

36. R. Diehl, T. Siegert, W. Hillebrandt, et al., SN2014J gamma rays from the 56Ni decay chain. Astron. Astrophys. **574**, A72 (2015)

37. R. Diehl, Gamma rays from a supernova of type Ia: SN2014J. Astron. Nachr. **336**(5), 464 (2015)

38. E. Churazov, R. Sunyaev, J. Isern, et al., Gamma-rays from Type Ia Supernova SN2014J. Astrophys. J. **812**(1), 62 (2015)

39. J. Isern, P. Jean, E. Bravo, et al., Gamma-ray emission from SN2014J near maximum optical light. Astron. Astrophys. **588**, A67 (2016)

40. M. Renaud, J. Vink, A. Decourchelle, et al., The signature of ^{44}Ti in Cassiopeia A Revealed by IBIS/ISGRI on INTEGRAL". Astrophys. J. Lett. **647**(1), L41–L44 (2006)

41. T. Siegert, R. Diehl, M.G.H. Krause, et al., Revisiting INTEGRAL/SPI observations of 44Ti from Cassiopeia A. Astron. Astrophys. **579**, A124 (2015)

42. L.G. Spitler, P. Scholz, J.W.T. Hessels, et al., A repeating fast radio burst. Nature **531**(7593), 202–205 (2016)

43. Chime/Frb Collaboration, M. Amiri, B.C. Andersen, et al., Periodic activity from a fast radio burst source. Nature **582**(7812), 351–355 (2020)

44. E. Fonseca, B.C. Andersen, M. Bhardwaj, et al., Nine new repeating fast radio burst sources from CHIME/FRB. Astrophys. J. Lett. **891**(1), L6 (2020)

45. M. Pilia, M. Burgay, A. Possenti, et al., The lowest-frequency fast radio bursts: sardinia radio telescope detection of the periodic FRB 180916 at 328 MHz. Astrophys. J. Lett. **896**(2), L40 (2020)

46. C. Gouiffes, L. Spitler, I. Cognard, et al., INTEGRAL and radio joint programme of FRB121102 during a renewed activity. The Astronomer's Telegram **13073**, 1 (2019)

47. D. Svinkin, D. Frederiks, K. Hurley, et al., A bright γ-ray flare interpreted as a giant magnetar flare in NGC 253. Nature **589**(7841), 211–213 (2021)

48. R. Margutti, B.D. Metzger, R. Chornock, et al., An embedded X-Ray source shines through the aspherical AT 2018cow: revealing the inner workings of the most luminous fast-evolving optical transients. Astrophys. J. **872**(1), 18 (2019)

49. B. Trakhtenbrot, I. Arcavi, C. Ricci, et al., A new class of flares from accreting supermassive black holes. Nat. Astron. **3**, 242–250 (2019)

50. C. Ferrigno, V. Savchenko, E. Bozzo, et al., INTEGRAL observation of the nuclear transient AT2019pev and the TDE candidate AT2019qiz. Astron. Telegram **13170**, 1 (2019)

51. A. Claret, P. Laurent, A. Sauvageon, et al., Dose measured on-board INTEGRAL after more than 12 years in space. IEEE Trans. Nucl. Sci. **62**(6), 2784–2791 (2015)
52. N.P. Meredith, R.B. Horne, I. Sandberg, et al., Extreme relativistic electron fluxes in the Earth's outer radiation belt: analysis of INTEGRAL IREM data. Space Weather **15**(7), 917–933 (2017)
53. M. Armano, H. Audley, J. Baird, et al., Measuring the galactic cosmic ray flux with the LISA pathfinder radiation monitor. Astropart. Phys. **98**, 28–37 (2018)
54. A. Struminsky, W. Gan, Observation of solar high energy gamma and X-ray emission and solar energetic particles. J. Phys. Conf. Ser. **632**, 012081 (2015)
55. M.K. Georgoulis, A. Papaioannou, I. Sandberg, et al., Analysis and interpretation of inner-heliospheric SEP events with the ESA standard radiation environment monitor (SREM) onboard the INTEGRAL and Rosetta missions. J. Space Weather Space Clim. **8**, A40 (2018)
56. J.U. Ness, A.N. Parmar, L.A. Valencic, et al., XMM-Newton publication statistics. Astron. Nachr. **335**(2), 210 (2014)
57. A.H. Rots, S.L. Winkelman, G.E. Becker, Chandra publication statistics. Publ. Astron. Soc. Pac. **124**(914), 391 (2012)
58. G.J. Babu, E.D. Feigelson, Goodness-of-fit and all that! ASP Conf. **351**, 127 (2006)
59. I.N. Reid, Gender-correlated systematics in HST proposal selection. Publ. Astron. Soc. Pac. **126**(944), 923 (2014)
60. G.J. Babu, E.D. Feigelson, Goodness-of-fit and all that! ASP Conf. **351**, 127 (2006)
61. F. Patat, Gender systematics in telescope time allocation at ESO. The Messenger **165**, 2–9 (2016)
62. C.J. Lonsdale, F.R. Schwab, G. Hunt, Gender-related systematics in the NRAO and ALMA proposal review processes (2016). arXiv e-prints
63. J. Carpenter, Systematics in the ALMA proposal review rankings. Publ. Astron. Soc. Pac. **132**(1008), 024503 (2020)
64. N. Reid, Personal communication (2023)

Chapter 6
Herschel Observing Time Proposals

Göran Pilbratt, Pedro García-Lario, and Arvind Parmar

Abstract After an introduction to the ESA Herschel Space Observatory including a mission overview, science objectives, results and productivity we examine the process and outcomes of the announcements of observing opportunities (AOs). For Herschel, in common with other ESA observatories, there were no rules, quotas, or guidelines for the allocation of observing time based on the geographical location of the lead proposer's institute, gender, or seniority ("academic age"); scientific excellence was the most important single factor. We investigate whether and how success rates vary with these ("other") parameters. Due to the relatively short operational duration of Herschel—compared to XMM-Newton and INTEGRAL—in addition to the pre-launch AO in 2007 there was just two further AOs, in 2010 and 2011. In order to extend the time-frame we compare results with those from the ESA Infrared Space Observatory (ISO) whose time allocation took place approximately 15 years earlier.

Keywords Herschel · Science results · Science productivity · Announcement of observing opportunities · Open time · Peer review · Time allocation committee · Panel members · Panel chairs · Observing priorities · Gender balance · User group · Proposers' institutes countries · Academic age · Proposal acceptance rates · Gender · Regional dependence · ISO comparison · Further activities

G. Pilbratt (✉)
Göteborg, Sweden
e-mail: glpil@pm.me

P. García-Lario
Directorate of Science, ESA/ESAC, Villanueva de la Cañada, Madrid, Spain
e-mail: Pedro.Garcia.Lario@esa.int

A. Parmar
Department of Space and Climate Physics, MSSL/UCL, Dorking, UK
e-mail: arvind.parmar@ucl.ac.uk

© The Author(s) 2025   155
A. Parmar et al., *ESA Science Programme Missions*, ISSI Scientific Report Series 18,
https://doi.org/10.1007/978-3-031-69004-4_6

6.1 Introduction

The Herschel Space Observatory was the European Space Agency (ESA) far-infrared and sub-millimetre space observatory cornerstone facility [1]. It was equipped with a large passively cooled telescope and a science payload consisting of three complementary focal plane instruments located inside a superfluid helium cryostat. During an almost four year period in 2009–2013 it opened up and exploited the then poorly utilised spectral range ~55–670 μm for imaging and spectroscopy of a wide variety of astronomical targets spanning the solar system to distant galaxies, even peaking into the first billion years after the Big Bang.

Herschel was the fourth and final cornerstone mission (CS4) of the Horizon 2000 Programme [2], following the already launched SOHO/Cluster (1995/2000) [3, 4], XMM-Newton (1999) [5], and Rosetta (2004) [6] missions. The Herschel science payload was selected in 1997–99 and the actual construction of the spacecraft by a large industrial consortium commenced in 2001.

A self-contained and much more complete description of Herschel, including chapters on its origin, history, science objectives and results, technology and technical innovations, and management, can be found in *Inventing a Space Mission: The Story of the Herschel Space Observatory* [7]. A nicely illustrated summary can be found in the *Herschel—Science and Legacy* brochure [8].

6.2 Mission Overview

Herschel was launched (together with the cosmic microwave background M3 mission Planck [9]) from Centre Spatial Guyanais (CSG) Kourou by an Ariane 5 ECA rocket on 14 May 2009. The V188 dual launch was the first of its kind and was carried out flawlessly, releasing Herschel about 26 minutes after liftoff (and Planck a couple of minutes later) perfectly on track as planned. Exactly a month later, on 14 June 2009 while en-route towards its operational large amplitude quasi-halo orbit around the L2 Lagrangian point and still in the process of cooling down, the cryostat cover was successfully opened (and locked open) by manual commanding. For the first time Herschel could see the sky and performed a "sneak preview" of the galaxy M51 through its 3.5 metre diameter passively cooled Cassegrain telescope [10], demonstrating good optical quality and correct alignment with focal plane units on the optical bench inside the cryostat. The telescope utilised (the then) novel silicon carbide technology and was the largest astronomical telescope ever operated in space until the launch of the James Webb Space Telescope (JWST) on 25 December 2021.

The science payload was designed to provide photometric imaging and spectroscopic capabilities over the entire spectral range. This demanded complementary instruments employing a number of different enabling technologies.

Two direct detection instruments each provided multi-band imaging photometry and low/medium spectral resolution imaging capabilities. The Photodetector Array Camera and Spectrometer (PACS) [11] covered the 55–210 µm range, providing either dual-band imaging over a 1.75' x 3.5' Field of View (FOV) or 5 x 5 pixel spectroscopy with a spectral resolution of between 1500 and 4000, while the Spectral and Photometric Imaging REciever (SPIRE) [12] covered the 194–672 µm range, providing either triple-band imaging over a 4' x 8' FOV or imaging spectroscopy with 2.6' diameter FOV and a spectral resolution of between 40 and 1000. Together PACS and SPIRE provided six wide photometry bands and low/medium spectroscopy covering the entire Herschel spectral range with a small overlap for cross-calibration.

The third instrument provided Herschel's very high spectral resolution capability. The Heterodyne Instrument for the Far Infrared (HIFI) [13] provided a single pixel on the sky with a bandwidth of up to 4 GHz anywhere between 480–1250 GHz (625–240 µm) or 1410–1910 GHz (212–157 µm) with a maximum spectral resolution beyond 10^6.

For photometry the observer could choose either of PACS or SPIRE, or a special "parallel mode" of operating both of these simultaneously, useful for covering large areas on the sky by scanning. For spectroscopy one of the three instruments could be operated in any one of a number of modes offered. It turned out that over the mission lifetime very similar amounts of observing time were spent on photometry and spectroscopy.

Herschel was an observatory operated as a partnership between ESA, the three instrument consortia and National Aeronautics and Space Administration (NASA). ESA was responsible for all mission operations, conducted through the Mission Operations Centre (MOC) at European Space Operations Centre (ESOC) in Darmstadt, Germany. The MOC was responsible for orbit maintenance, daily operations, uplink and downlink, including the health and safety of the spacecraft and instruments, supported as necessary by the other partners. For communication with the spacecraft either of ESA's New Norcia (Perth, Australia) or Cebreros (Avila, Spain) deep space antennas was used in a single scheduled 3-hour window every day.

The Herschel Science Ground Segment (SGS) [14] was a partnership consisting of five elements: the Herschel Science Centre (HSC) located at the European Space Astronomy Centre (ESAC) near Madrid, Spain, an Instrument Control Centre (ICC) for each Herschel instrument provided by its consortium based at Netherlands Institute for Space Research (SRON), Groningen, The Netherlands (HIFI), at Max Planck Institute for Extraterrestrial Physics (MPE), Garching, Germany (PACS), and at Rutherford Appleton Laboratory (RAL), Didcot, UK (SPIRE), and the NASA Herschel Science Center (NHSC) located at the Infrared Processing and Analysis Center (IPAC) in Pasadena, California, USA. The HSC performed scientific mission planning and was the prime interface between Herschel and the science community, supported by the partners. It provided information and user support related to the entire life-cycle of Herschel observations, including calls for observing time, the proposing procedure, proposal tracking, and supported the Herschel Observing

Time Allocation Committee (HOTAC) and the Herschel Users' Group (HUG) in executing their tasks. The HSC also performed systematic pipeline data processing and hosted and populated the Herschel Science Archive (HSA) through which access to the Herschel data was, and still is, offered to the worldwide astronomical community, and provided the Herschel Interactive Processing Environment (HIPE) for interactive data processing [15]. The building of the necessary extensive software systems, in particular HIPE [16], together comprising Herschel Common Science System (HCSS) was led by the HSC and depended critically on contributions from all members of the SGS.

The mission lifetime was limited by the continuously diminishing supply of the superfluid liquid helium cryogen which eventually was exhausted, with the last scientific observation being carried out on 29 April 2013. After some post-operations technical tests and eventually having been placed in its heliocentric "disposal orbit", the spacecraft was finally switched-off on 17 June 2013. Herschel has successfully conducted ~37,000 science observations recommended by the HOTAC in ~23,400 hours of observing time, and additionally ~6600 science calibration observations in another ~2600 hours, all of which, in the form of a variety of derived data products at different levels of processing, are available through the HSA.

6.3 Science Objectives and Results

Given the considerable time between the 1984 selection of Herschel—called the Far Infra-Red and Submillimetre space Telescope (FIRST) at the time—and implementation naturally its science objectives were repeatedly subject to review and presented to the community for feedback, in the course of multiple studies undertaken. Specifically, this was the case ahead of the finalisation of its Science Management Plan (SMP) and the Announcement of Opportunity (AO) for its science payload, both issued in the autumn of 1997, consisting of a special "hearing" with invited experts in September 1996 and the "Grenoble meeting" held in April 1997 [17].

Since Herschel was to "open up" a new (then) poorly observed part of the spectral regime, "the cool universe" (the blackbody peak corresponding to the spectral coverage is in the range 5–55 K), the idea was that the mission needed both to survey and follow-up on its own, while at the same time being lifetime constrained. Therefore the SMP required that large observing programmes, called Key Programmes (KPs), should be selected and conducted early upfront in the mission. Special efforts were made to inform and engage the community early on, starting in earnest with the "Toledo conference" held in December 2000 [18], in order to provide the community with ample time to prepare scientifically and organisationally to respond to the Key Programme (KP) AO. This was also the meeting when Herschel got its name.

Main Science Objectives The top-level scientific areas foreseen for Herschel included wide-area galactic and extragalactic surveys, pertaining in particular to dust-enshrouded star formation throughout cosmic time; detailed studies of the Interstellar Medium (ISM) in the Galaxy and other, primarily nearby resolvable, galaxies; observational astrochemistry of gas and dust in a variety of objects, as a tool for understanding physical and chemical processes throughout stellar lifecycles; and investigation of a wide variety of Solar System objects and their atmospheres. Herschel has delivered on all accounts, what follows is a—most likely but unintentionally biased—selection displaying the wide scope of scientific results.

Extragalactic Surveys Two major photometric survey KPs were conducted, supplemented by a number of other programmes. The number of "sub-millimetre" galaxies catalogued has been increased from some hundreds, or at most a couple of thousand, to in excess of half a million [19, 20], corresponding typically to 600+ sources per square degree on the sky. Herschel data suggest that although at all cosmic epochs the most vigorously star forming galaxies seem to be interacting galaxies [21, 22], at least for redshifts up to $z\sim2.5$ (corresponding to a look back time of ~11 Gyr) most of the star formation occurs in secularly star-forming galaxies where the infrared luminosity increases with redshift [23]; sometimes referred to as the "galaxy main sequence", thus what a "normal" star-forming galaxy is depends on cosmic epoch. The generally greater productivity at earlier epochs appears simply to be associated with the availability of cold molecular gas— the raw material for making stars inferred from the thermal emission of dust that Herschel has observed—which was more plentiful at earlier epochs [24–26]. As a population the galaxies detected in Herschel deep surveys resolve the cosmic infrared extragalactic background (which peaks at around 160 μm) into discrete sources [27], removing any need for more imaginative contributions.

Early Universe Herschel has also detected some galaxies in the first billion years (redshifts of $z\simeq6$ or greater) after the Big Bang [28], and many more in the first couple of billion years [29–31]. These detections require massively star forming galaxies, with rates of thousands of solar masses per year (currently in the Galaxy the rate is about one solar mass per year). With the reservation that (a poorly known fraction of) these could be spatially unresolved groups of a few galaxies, the current understanding of structure formation and galaxy evolution in the very early universe is stretched.

Ultra Luminous InfraRed Galaxies (ULIRGs) Spectral studies of relatively "local" ULIRGs, essentially all of which are mergers of galaxies, have shown that the vast majority, if not all, of them display massive molecular outflows emanating away from them. These winds with velocities up to in excess of 1000 km s^{-1} [32, 33] are so powerful that they remove molecular material much faster than consumed by star formation. Unchecked they could remove all material to form any new stars in just millions or tens of million years, very short timescales on galaxy scales; an extreme form of "feedback" [34] indeed.

Supernova Remnants A chance Herschel discovery was the remnant of a recent supernova, SN1987A, an unexpected by-product of a photometric survey of the Large Magellanic Cloud, a satellite galaxy of the Galaxy. The measured fluxes implied that enormous amounts of dust had been created in the explosion [35]. If this amount of dust is formed by a typical supernova it would explain the amount of dust that Herschel has observed in the early universe. However, there is great uncertainty about how much dust actually survives from the "reverse shock" which is generated by the outgoing shock and travelling back towards the site of the explosion. This is a very interesting result, but the potential implications are not yet fully clear [36]. Another serendipitous supernova related detection was that of the argon hydride ion $^{36}ArH^+$ through its J=1-0 and 2-1 lines in the Crab Nebula [37]. The ^{36}Ar isotope is expected from the explosive nucleosynthesis in a core collapse supernova of a massive star, thus providing direct support for the Crab Nebula as a remnant of the supernova observed by the Chinese in 1054.

Low Mass Star Formation Giant molecular clouds of interstellar matter near and far have been surveyed in multiple bands. Studies of low mass star formation in the nearby regions in the Gould Belt [38] show that these display intricate networks of filamentary structures [39] with characteristic widths [40, 41]. Herschel found them in all regions, some of which were forming stars [42–44], but not all [45]. A closer analysis [46] suggests that turbulence creates the structures, but that they need to be massive enough to become gravitationally unstable and fragment to form stars; Herschel data have enabled numbers to be assigned on what "massive enough" appears to be [47, 48] and have generated speculations about a new paradigm of (low mass) star formation [49].

High Mass Star Formation A complementary survey covered essentially all regions forming massive, OB, stars at distances less than 3 kpc from the sun, creating spectacular images [50, 51]. A 1.2 THz wide HIFI spectral survey in Orion, the closest massive star formation region, was conducted as part of a larger programme [52] and revealed a total of ~13,000 lines from 79 isotopologues of 39 different molecules, with excellent agreement between observations and modelling [53] achieved. The chemical modelling [54] will represent a legacy for comparison with other chemical models and sources.

Galactic Plane Surveys In addition to specific galactic regions a photometric mapping of the entire 360° of the galactic plane with a latitude coverage of ±1° has been conducted in about 1000 hours of observing time [55, 56]; this is the largest chunk of Herschel observing time allocated to a single purpose. The survey takes the view of the Galaxy as a star formation engine [57], attempts are made to understand star formation globally in the Galaxy, cataloguing several hundred thousand compact sources [58–60]. Using this survey data the overall structure of the Galaxy [61] and the large-scale properties of its interstellar medium have also been studied [62]. Spectrally a major survey of the 158 μm ionised carbon, C^+, line along over 500 lines of sight through the galactic plane has been conducted [63]. It has shown that there is significantly more galactic molecular hydrogen gas than currently inferred from carbon monoxide observations, the so-called "CO-dark" component [64, 65]. The C^+ line is also potentially a powerful star formation tracer [66].

Infrared Excess Stars Herschel was not particularly suited to observe "normal"—main sequence—stars, like the sun, and was not normally used to, but an exception is the study of "infrared excess" which originates from dust surrounding a star (the "Vega phenomenon"). Two KPs were conducted, [67, 68], together observing hundreds of nearby stars, finding and characterising infrared excess ("exo-Kuiper belts") to new levels of precision and resolution [69–71], resulting in incidence levels for nearby FGK stars of $\simeq 21\%$ [72, 73]. The nearby Fomalhaut displays a spectacular disc [74] and in the young β Pic the 69 μm band of crystalline olivine has been detected, exhibiting strikingly similar solid state properties to those of the dust emitted from the most primitive comets in the solar system [75].

Late Type Stars Stars in the process of formation accrete material while late-type ("dying") stars shed material, interacting with their surroundings in both cases. The morphology of the inner envelope and bow shock around the red supergiant Betelgeuse (α Ori) was imaged in all six Herschel bands [76], displaying complicated multi-arch features in the bow shock region. A spectral survey was conducted on IRC+10216 (CW Leo). It showed a plethora of lines, many not previously observed, from many different molecules [77]. Unexpectedly—since for a "carbon-rich" star the oxygen was expected to be locked up mainly in CO and SiO—no less than 39 ortho-H_2O and 22 para-H_2O lines, including low- and high-excitation lines from different parts of the envelope, were observed. Herschel has enabled important diagnostics of physical conditions and dynamics, for instance enabling to trace the mass loss of history of evolved stars, furthering the understanding of late stages of stellar evolution and the chemical enrichment of the interstellar medium [78].

Water "Trail" Water is of particular interest, as a diagnostic line, but also as a coolant, and in its own right for its connection to life as we know it. Consisting of the two most common reactive elements (hydrogen and oxygen) in the universe, it forms on the surfaces of grains in molecular clouds, the birth places of stars. Thanks to its sensitivity, high spectral resolution, and capability to observe the water ground level transitions Herschel has managed to study the "water trail" from formation to planets [79]. Herschel has detected water for the first time in a pre-stellar core in L1544 [80], displaying an inverse P-Cyg profile characteristic of contraction. In star-forming regions water is a key tracer of dynamics and chemistry, with complicated line profiles indicative of both out- and inflows and shocks, but are typically dominated by broad features indicating bulk outflows [81]. Herschel has provided the first observations of cold water vapour emission in the disc around the young star TW Hya [82], the vapour is in equilibrium with solid ice. The line was weaker than expected, but has enabled modelling of total water (vapour plus ice, where almost all the mass is in the ice) content. Later, the hydrogen deuteride (HD) fundamental (J=1-0) line was detected [83], enabling a more direct modelling of the total disc mass, which is enough to form a planetary system like that of the solar system; despite having a relatively advanced, albeit still uncertain, age.

Water in the Solar System Water is an important constituent in comets, but has also been detected by the Infrared Space Observatory (ISO) in 1997 in the stratospheres of all the four giant planets; a surprise discovery at the time the origin has been debated ever since. Herschel observations link the water observed in Jupiter's upper atmosphere to most likely originating from the 1994 impact of comet Shoemaker-Levy 9 [84], thus demonstrating water can be delivered by cometary impacts, and in Saturn's atmosphere originating from geysers on its moon Enceladus [85, 86]. Furthermore, Herschel has made the first direct detection of water vapour around the dwarf planet Ceres [87] in the main asteroid belt, and thus inside the "snowline".

Origin of Water on Earth Comets were important Herschel targets. Early on water in the comet C/2008 Q3 Garradd [88] was detected. For comets a particular interest was to measure the D/H-ratio of their water in order to compare with that of the oceans of the Earth, since the origin of the water on Earth is an open question subject to dispute. Before Herschel only Ivuna-type carbonaceous (CI) chondrites had measured D/H ratios similar to that of Earth's water. However, Herschel measured the same ratio for the Jupiter family comet 103P/Hartley 2 [89], which "made a splash" at the time. Then a higher ratio for the Oort cloud comet Garradd (in line with expectations for Oort cloud comets) was obtained [90], and yet later an upper limit, consistent with the Earth/Hartley 2 value, for the Jupiter family comet 45P/HMP [91]. However, in 2014 Rosetta measured a very much higher ratio—higher even than for the Oort cloud comets—for its target, the Jupiter family comet 67P/CG [92]. Not only is the issue of the provenance of Earth's water not resolved, but that of the D/H-ratio of comets has become more complicated than the pre-Herschel assumed simple picture.

Trans-Neptunian Object (TNO)s Herschel has also observed more than 130 of the approximately 1500 known TNOs, the largest being Pluto and Eris, with >90% detected for which photometric flux values have been obtained. These observations have transformed these cold "objects" outside the orbit of Neptune, to amazingly diverse "worlds" [93–96]. A few of the larger ones appear to have high albedos, indicative of "fresh" surfaces; could these be geologically active and able to resurface themselves, rather than being dead remnants of the formation of the solar system 4.5 billion years ago? The observations of Pluto made by NASA's New Horizons in 2015 may indicate this is true, at least for this particular TNO.

ESA "Web Releases" Most if not all of the above examples, and in fact many more, have been the subject of various ESA "web releases"; for a listing of all of these with links to stories and sources as appropriate in each case consult: https://www.cosmos.esa.int/web/herschel/press-releases and https://www.cosmos.esa.int/web/herschel/latest-news.

6.4   Science Productivity

The wide scope of Herschel's science objectives and its spectral coverage made for a large user community, embracing the ISO and Spitzer communities (the NASA mission Spitzer ran out of cryogen the day after Herschel was launched, a remarkable coincidence!) as well as radio astronomers including the Submillimeter Wave Astronomy Satellite (SWAS) and Odin communities. This community made use of the data, produced science and published papers. Herschel also naturally has provided complementary views, to in particular radio, near infrared/optical and X-ray observations, some of them sometimes considered controversial from the perspectives of these other communities. Herschel has also complemented the Planck all-sky survey of the cosmic microwave background by providing higher angular resolution and spectral capabilities to "zoom in" on particular sources, doing astronomy and potentially aiding in the crucial removal the Planck "foregrounds".

Given the limited lifetime of the mission special efforts were made early during the in-flight mission to keep the community informed and updated of the actual demonstrated observatory capabilities, in particular in preparation for the first in-flight AO, the AO1 (see below) that took place just within a year after the launch. Small numbers of observations from almost all KPs were selected for early observing and release in what was called the Science Demonstration Phase (SDP). A number of meetings were organised, culminating in early May 2010 with the "Herschel First Results Symposium", immediately followed by pre-prints of accepted papers which were placed on the arXiv open-access archive. These were then published in two Special Issues of *Astronomy & Astrophysics* dedicated to Herschel [97, 98], and the scientific data were made publicly available to the entire community through the HSA.

This effort generated 202 papers in the *A&A* Special Issues in 2010 alone, a good start that was further improved upon in the coming time and years. In fact, in the 10 calendar years following the launch year, Herschel has the highest number of publications of any of the ESA-led observatories to date, cf. Fig. 6.1. This is the more noteworthy given that by less than 4 years after launch Herschel was no longer observing (as was true for ISO about 2.5 years after its launch) while both XMM-Newton and INTEGRAL are still active space observatories more than 20 years after their respective launches. Interestingly—but outside the time range of the figure— the launch of Herschel appears to have stimulated the use of ISO data as the rate of ISO papers published increased at the time.

Up to the end of 2021, there have been almost 3300 refereed Herschel publications (cf. Table 2.1 and Fig. 2.1) that meet the inclusion criteria given in Chap. 2. In addition to a healthy publication record Herschel has also consistently scored well with respect to various other related parameters (cf. Fig. 2.24), the number of citations is high (cf. Fig. 2.28) with a considerable number (cf. Fig. 2.27) of i100 papers (papers with more than 100 citations) and a high h-index (cf. Fig. 2.29). Herschel also has been scoring well in the ESA defined Key Performance Indicator (KPI)s (cf. Fig. 2.30 for one of them) regularly reported to the Science Programme

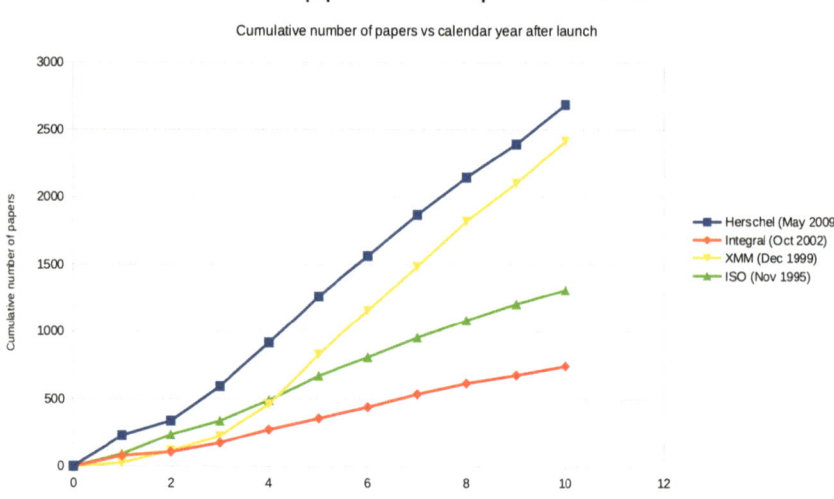

Fig. 6.1 Cumulative numbers of refereed papers for the four ESA-led space observatories ISO, XMM-Newton, INTEGRAL, and Herschel in the ten calendar years following their respective launch years (year 0 in the figure). ISO terminated observing in year 3 and Herschel in year 4 (in both cases due to the exhaustion of cryogen), while XMM-Newton and INTEGRAL are still observing more than 20 years after launch

Committee (SPC), and has generated a large number of Doctor of Philosophy Degree (PhD) theses over the years, mainly related to its science but also with a fraction related to its technical development.

Given in particular the KPs some Herschel programmes, or combinations of programmes, produced large data sets that were considered to potentially be of great legacy value. This generated a desire to produce suitable legacy datasets to made publicly available for the entire community, enhancing the value of the Herschel data and enable additional science to be performed. This would sometimes involve additional datasets and custom data processing and tools, requiring resources not available within the Herschel consortia on their own.

The European Union 7th Framework Programme (FP7) offered a potential solution by making funds available through competitive applications. Herschel teams were extremely successful, the two proposals HELP—"Herschel Extragalactic Legacy Project" [99] and VIALACTEA—"The Milky Way as a Star Formation Engine" [100] each obtained the full score and were each awarded with 2.5 M€. In addition, DustPedia—"A Definitive Study of Cosmic Dust in the Local Universe" [101] was awarded another 2 M€ from FP7, and later SBNAF—"Small Bodies Near and Far" [102] was awarded 1.5 M€ from the follow-up EU Horizon 2020 programme. It is, however, unclear just exactly to what extent these projects have been fully successful in achieving their very ambitious goals.

6.5   Observing Time

Herschel was designed to provide just under 20,000 hours for allocation to conduct scientific observations, out of which 32% was Guaranteed Time (GT), mainly for the instrument consortia reflecting the requirements put on them to contribute to the science operations throughout the mission lifetime. The remainder was Open Time (OT) which was open to the worldwide general astronomical community through a standard competitive AO observing proposal submission process.

The KPs were to be selected through a dedicated initial AO before the launch, by necessity since these would be the first observing programmes to be conducted. This AO thus would have to cope with the additional difficulty of an imprecise knowledge of the scientific capabilities of the observatory. To scientifically "validate" the selected KPs given the actual performance before commencing full-scale observing was one of the reasons for the introduction of the SDP. Additional AOs would follow later in-flight.

The first Herschel AO, the dedicated "KP only AO", for proposals requiring substantial amounts (>100 hours) of observing time, was released in February 2007. Like all Herschel AOs it consisted of two parts; first a phase for Key Programme Guaranteed Time (KPGT), followed by a phase for Key Programme Open Time (KPOT) after the accepted GT programmes and associated observations had been identified and communicated. KPOT proposals could only be submitted after the conclusion of the KPGT phase. The entire KP AO process was completed in February 2008, with the publication of the successful KPOT programmes and observations. By coincidence it resulted in 21 KPGT and 21 KPOT proposals being awarded observing time. The first in-flight announcement (AO1) took place in 2010 with the initial guaranteed and subsequent open time phases being labelled GT1 and OT1. In 2011, the second, and final, announcement (AO2) was released with GT2 and OT2 phases as before; see Table 6.1 for an overview.

For the in-flight AO OT phases the concept of priority 1 and priority 2 observing time was introduced. The background was that it had become clear that at all times it was necessary to have a pool of (at least) 6 months of observations in order to enable efficient scheduling. Ultimately this would mean that when Herschel ran out of helium about 6 months of accepted observations would remain unobserved, they would not be performed. A connected and important problem here, in particular regarding OT2, was a large uncertainty in the predicted lifetime. To cut a complicated story short, in the end approximately half of the priority 2 observations were actually executed. These were selected based primarily on grading (essentially all of the top third graded priority 2 proposals were observed, about half of the middle tier, and only a small fraction of the lower tier) but necessarily also on various additional constraints, in particular with respect to sky visibility which changed throughout the year.

In addition, for completeness, but for the purpose of this study negligible amounts of observing time were awarded as Director's Discretionary Time (DDT), "Must-Do", and "Filler" programmes.

Table 6.1 Herschel AO overview showing dates, number of proposals, and awarded observing time. For completeness here both GT and OT are shown. The meaning of priorities p1 and p2 are described in the text. For the KP and GT AOs the concept of priorities was not applicable, the awarded time is equivalent to p1

AO phase	Date of		Number of proposals				Time (hour)		
	issue	Deadline	Sub-mitted	Accepted			Awarded		
				Total	p1	p2	Total	p1	p2
KPGT	Feb 2007	Apr 2007	21	21	21	...	5879	5879	...
KPOT	Feb 2007	Oct 2007	62	21	21	...	5379	5379	...
GT1	Feb 2010	Mar 2010	33	33	33	...	555	555	...
OT1	May 2010	Jul 2010	577	241	176	65	6577	4987	1590
GT2	Apr 2011	May 2011	32	32	32	...	362	362	...
OT2	Jun 2011	Sep 2011	531	373	181	192	7795	3420	4170

6.6 Herschel Observing Time Allocation Committee (HOTAC)

The HOTAC was an external committee that evaluated proposals and made recommendations on the allocation of observing time to ESA Director of Science (D-SCI). Since the operational lifetime of Herschel crucially depended on its cryogen supply it had a limited, and for a major observatory relatively short, lifetime. The HOTAC was formed with the hope that it could serve for the entire mission with a minimum of changes. It was decided that the GT proposals should also be reviewed by the HOTAC, for two reasons—all proposals should represent science worthy of the limited availability of precious Herschel observing time, and by reviewing the GT proposals the HOTAC would obtain a good overview of, and familiarity with, the already allocated observing time in the various science areas, which was deemed useful when reviewing the OT proposals.

For operational reasons, and given the two (GT and OT) phases of each cycle, it was concluded that only one AO cycle per year could be supported, both from the HOTAC as well as from the conducting the AO, importantly including providing the necessary support to the HOTAC, activities. For the OT phases the task of the HOTAC was to grade the submitted proposals taking into account the scientific case, justification, merit and relevance of the proposed observation(s), and the potential contribution of the overall scientific return of the mission. The HOTAC then made recommendations on the allocation of observing time, which could include modifications to what the proposer requested. The final decision concerning observing time allocation was performed by the D-SCI.

The HOTAC consisted of an overall chair and initially two parallel panels covering each of four science areas Interstellar Medium, Star Formation and Solar System Objects (ISM/SF/SS), Stars and Stellar Evolution (S/SE), Galaxies and Active Galactic Nuclei (G/AGN) and Cosmology. Each panel worked independently, and then a meeting at what became to be referred to as the "HOTAC Level", consisting of

Table 6.2 A summary of the HOTAC membership. The HOTAC Chair is included in both the Full HOTAC and the HOTAC Level. The HOTAC Level members are included in the Full HOTAC. Although numbers appear persistent, actual compositions were not, with the exception of the Chair who was the same individual throughout (see text). %f is the female percentage

AO	Year issued	No. of proposals	Full HOTAC			HOTAC level			HOTAC chair	
			Male	Female	%f	Male	Female	%f	Male	Female
KPOT	2007	62	–	–	–	9	4	31	0	1
OT1	2010	577	29	13	31	9	4	31	0	1
OT2	2011	531	29	13	31	9	4	31	0	1

the overall chair and the panel chairs and vice-chairs, would derive the final overall HOTAC recommendation. The HOTAC chair would then communicate this to the D-SCI. Since in the KP AO only a small number of proposals were expected, but collectively they would encompass a large amount of observing time, it was decided that for this AO all the work with all the proposals would be carried out at the HOTAC level, consisting of 13 people, thus panel members were not recruited at this point.

For the AO1 and AO2 cycles, based on the KP AO proposal response in the various science areas, the HOTAC structure was enlarged with a third ISM/SF/SS panel. The ISM/SF/SS and G/AGN panels each had five members, and each of these two science areas supplied four people to the HOTAC Level, while S/SE and Cosmology panels each had four members, and each of these science areas supplied two people to the HOTAC Level; these numbers approximately mirroring the proposal pressure. Thus, the Full HOTAC consisted of 42 people, the HOTAC Level of 13 people, in both cases including the overall Chair, see Table 6.2. Although the objective was to keep as many people as possible from each AO cycle to the next, each time approximately 15% of the members were replaced, however, the HOTAC Chair remained, and magically so did the gender distribution.

6.7   Herschel Users' Group (HUG)

The HUG had in common with the HOTAC that it was an independent advisory group that was self-managed and was not a part of the Herschel SGS (which had its own internal advisory groups). It was set up on the initiative of the Herschel Science Team (it was actually not required by, or even mentioned in, the SMP) in the year of the launch, to provide a forum for (potential) general users of Herschel to provide input to the operations of the observatory as an astronomical facility on various matters affecting its scientific productivity and user friendliness.

The HUG conducted a preparatory teleconference in addition to a total of ten meetings in the period 2010–2016. It interacted directly with the Herschel user community, e.g., through polls and their own networks, it was supported by the Project Scientist (PS) and the SGS. It reported to the PS who posted all the

Table 6.3 A summary of the aggregate HUG membership, its composition varied over time (see text). The HUG Chair is included in the Full HUG, six out of the ten meetings held had a female chair. %f is the female percentage

Full HUG			HUG Members			HUG Chairs		
Male	Female	%f	Male	Female	%f	Male	Female	%f
10	7	41	9	6	40	1	1	50

presentations made in the meetings and the minutes with the HUG advice on a dedicated HUG web-page publicly available for everyone, and brought the HUG recommendations to the attention of the Herschel Science Team and the SGS.

Thus the HUG was "by and for" the user community. The original membership was eight people drawn from the already existing KPOT user community, later extended by four additional members drawn from the OT1 community. HUG members served for two years, and thus the membership was updated on a regular basis throughout, also including members drawn from the OT2 community. NASA had already somewhat earlier put in place their own NHSC Users' Panel (NUP), and the two committees not only collaborated by also had overlapping membership, the NUP chair was a member of the HUG. A summary of the HUG membership can be found in Table 6.3.

6.8 Proposers

The OT phases of all three Herschel AOs were open to the worldwide community in a similar manner to AOs from other ESA missions. There were no rules applicable to nationalities, and there were also no quotas or other kinds of limitations or guidelines in place with respect to the allocation of the available observing time. When submitting a proposal in response to an AO a single Principal Investigator (PI) had to be named in addition to all the other co-investigators, all with affiliations. From the perspective of ESA the proposal PI was the contact point for the proposal for all kinds of communication, e.g., the allocation of time and information provision. The information provided by the PIs allows an examination of the nationality distribution of the proposers' institutes. Table 6.4 shows the number of proposals submitted and accepted covering all 1170 proposals. Accepted proposals are ones which were awarded any observing time at either priority, i.e. both p1 and p2 (cf. Sect. 6.5 Observing Time).

It is immediately noticeable that just over 40% of all submitted proposals were from PIs located in institutes in the USA. This is slightly more than the sum of the five ESA Member States with the largest numbers of proposals submitted; the same is true when it comes to numbers of accepted proposals. This illustrates one effect of having a fully open proposal procedures, another effect is the science return in terms of refereed publications (Chap. 2).

Table 6.4 The countries where the Herschel PIs were located. The KPOT, OT1 and OT2 calls resulted in a total of 1170 submitted proposals with 644 (55.0%) being accepted (allocated any observing time). The PIs were located at institutes in the following countries

Country	No. proposals submitted	Percentage of total	No. accepted	Percentage accepted
BELGIUM	13	1.1	7	53.8
DENMARK	5	0.4	1	20.0
FINLAND	3	0.3	2	66.7
FRANCE	90	7.7	58	64.4
GERMANY	101	8.6	49	48.5
GREECE	3	0.3	1	33.3
HUNGARY	15	1.3	7	46.7
IRELAND	1	0.1	1	100
ITALY	60	5.1	29	48.3
NETHERLANDS	69	5.9	41	59.4
POLAND	3	0.3	2	66.7
PORTUGAL	2	0.2	2	100
SPAIN	75	6.4	35	46.7
SWEDEN	15	1.3	8	53.3
SWITZERLAND	10	0.9	5	50.0
UNITED KINGDOM	132	11.3	75	56.8
ARMENIA	2	0.2	1	50.0
AUSTRALIA	11	0.9	7	63.6
CANADA	31	2.6	18	58.1
CHILE	1	0.1	1	100
CHINA	2	0.2	2	100
INDIA	5	0.4	2	40.0
ISRAEL	4	0.3	4	100
JAPAN	29	2.5	12	41.4
RUSSIAN FEDERATION	2	0.2	0	0.0
SLOVAKIA	1	0.1	0	0.0
SOUTH AFRICA	2	0.2	1	50.0
SOUTH KOREA	4	0.3	1	25.0
TAIWAN	1	0.1	1	100
UNITED STATES	478	40.9	271	56.7

6.9   Proposal Selection

We examined the outcomes in terms of success based on gender for the OT announcements (KPOT, OT1 and OT2) which, as discussed in Sect. 6.8 were fully open to the worldwide community, without restrictions of any kind. Of the 1170 proposals submitted in response to these AOs, 644 (55.0%) of the them were awarded *any* observing time, i.e. p1 or p2 time, and not necessarily all the time requested.

In order to investigate success rates of proposals for male and female PIs it was necessary to assign genders to the proposers since proposers were not asked to specify their gender (nor age, nor type of position held etc.). As with XMM-Newton and INTEGRAL, gender information for each proposer was assigned by the project scientist and Science Operations Centre (SOC) staff using their knowledge of the community and through examining publicly-accessible web-based data in a similar way as for the Hubble Space Telescope (HST) study by I.N. Reid [103]. We appreciate that gender identity is more complex than a binary issue, however, no attempt can be made to assign genders other than male or female as this information is not readily available. It is possible that this process underestimates the number of female proposers, particularly those with non-Western names. However, this is likely be insignificant given the small fraction of non-Western proposers.

We first examined the number of proposals that were awarded any observing time. As listed in Table 6.5 828 proposals with male PIs and 342 proposals with female PIs were submitted from which 461 and 183 proposals were awarded any observing time. This gives success rates of 55.7 and 53.5% for proposals led by male and female PIs, respectively. This is a difference in favour of male PIs of 4.1%.

Proposal submissions and proposal acceptance are unlikely to be independent or random processes and for Poisson statistics to apply events need to be independent

Table 6.5 Herschel OT proposals submitted with breakdown of proposals with male and female PIs and the number and percentages of proposals accepted as well as time and percentages awarded. KPOT proposals were not given priorities and all accepted proposals were counted as being priority p1

Priority	Parameter	Proposal PI		
		All	Male	Female
	Proposals submitted	1170	828	342
p1, p2	Number accepted	644	461	183
	Percentage accepted	55.0	55.7	53.5
	Ratio male/female		1.041	
p1	Number accepted	386	281	105
	Percentage accepted	33.0	33.9	30.7
	Ratio male/female		1.105	
	Time requested (hour)	51,422	39,996	11,446
p1, p2	Time accepted (hour)	19,751	15,336	4415
	Time accepted (%)	38.4	38.3	38.6
	Ratio male/female		0.994	
p1	Time accepted (hour)	13,934	11,464	2469
	Time accepted (%)	27.1	28.7	21.6
	Ratio male/female		1.328	
	Overall male/female difference		−1–33%	

of each other. However, to illustrate the uncertainties that would apply if such statistics are applicable, we used square root uncertainties to update the success rates to be (55.7 ± 3.2) and $(53.5 \pm 4.9)\%$ for male and female PI proposals, respectively. This is a difference in favour of males of $(4.1 \pm 11.3)\%$. We emphasise that these uncertainties are only given to illustrate the outcomes if Poisson statistics were to apply to the selection process. We note that actual uncertainties on the proposal numbers for each AO are zero. We further note that there are systematic uncertainties associated with this process due to misappropriated genders and incorrect PhD "academic ages", but these are likely to be too small to significantly affect our results.

We examined the number of AO1 and AO2 proposals that were awarded p1 observing time (KPOT proposals are counted as p1). As also listed in Table 6.5 of the 828 proposals with male PIs and 342 proposals with female PIs that were submitted, 281 and 105 proposals were awarded p1 observing time. This gives success rates of 33.9 and 30.7% for proposals led by male and female PIs, respectively. This is a difference in favour of male PIs of 10.5%.

We next examined the acceptance rates for the two genders using the awarded amounts of observing time. Many accepted Herschel proposals were allocated less observing time than requested. If the allocation of observing time was handled differently for male and female PIs, this will show as a difference in the relative amounts of time approved, compared to the number of proposals accepted. The results for each AO are summarised in Table 6.6. This shows the total times requested by proposals with male and female PIs were 39,996 hour and 11,446 hour, respectively. A total of 15,336 hour and 4415 hour of p1+p2 observing time were awarded to proposals with male and female PIs, respectively. This gives average success rates of 38.3% for male PIs and 38.6% for female PIs. This is almost equal with a small ($<1\%$) difference in favour of female PIs. However, when only p1 time is considered a total of 11,464 hour and 2469 hour observing time were awarded to proposals with male and female PIs, respectively. This gives average success rates of 28.7% for male PIs and 21.6% for female PIs, which amounts to a difference of 32.8% in favour of male PIs. Initially we were intrigued by this situation, given all observing time there is negligible difference, but for p1 time only the situation is different. We investigated possible explanations for this. An inspection of Table 6.6

Table 6.6 The requested and allocated observing time for male and female Herschel proposers. The last two columns provide the percentage of the allocated as a fraction of the requested observing time for the genders

| | Time (hours) | | | | | | Time (Percentage) | | | |
| | Requested | | Allocated (p1, p2) | | Allocated (p1) | | Allocated (p1, p2) | | Allocated (p1) | |
AO	Male	Female	Male	Female	Male	Female	Male	Female	Male	Female
KPOT	15,729	2255	5001	378	5001	378	31.8	16.8	31.8	16.8
OT1	15,149	5823	4776	1801	3739	1238	31.5	30.9	24.7	21.3
OT2	9118	3368	5559	2236	2725	853	61.0	66.4	29.9	25.3
Total	39,996	11,446	15,336	4415	11,464	2469	38.4	38.7	28.7	21.6

Table 6.7 Similar to Table 6.5 but providing the same information for KPOT and OT1+OT2 proposals separately in order to display the large influence of the small in number but large in time KPOT proposals. See discussion in the text

Priority	Parameter	Proposal PI					
		OT1+OT2	Male	Female	KPOT	Male	Female
	Proposals submitted	1108	774	334	62	54	8
p1+p2	Number accepted	623	442	181	…	…	…
	Percentage accepted	56.2	57.1	54.2	…	…	…
	Ratio male/female		1.054			…	
p1	Number accepted	386	281	105	21	19	2
	Percentage accepted	34.8	36.3	31.4	33.9	35.2	25.0
	Ratio male/female		1.155			1.407	
	Time requested (hour)	33,458	24,267	9191	17,984	15,729	2255
p1+p2	Time accepted (hour)	14,372	10,335	4037	…	…	…
	Time accepted (%)	43.0	42.6	43.9	…	…	…
	Ratio male/female		0.970			…	
p1	Time accepted (hour)	8555	6464	2091	5379	5001	378
	Time accepted (%)	25.7	26.8	22.8	29.9	31.8	16.8
	Ratio male/female		1.176			1.896	
	Overall male/female difference		−3–18%			41–90%	

indicates that the KPOT numbers stand out. In order to investigate further Table 6.7 was constructed, separating the KPOT results from the total (All) numbers, and comparing them directly with OT1+OT2 only (i.e. All − KPOT).

Inspection of Table 6.7 reveals that for OT1+OT2 male PIs have an advantage in number of accepted proposals of $(5.4 \pm 1.6)\%$, but were awarded 3% less p1+p2 observing time. As above, we used square root uncertainties to illustrate the uncertainties that would apply if such statistics are applicable. When OT1+OT2 proposals that were allocated p1 observing time are considered, male PIs have an advantage in numbers of $(15.5 \pm 15.2)\%$ and in allocated time of 17.6%. However, for the KPOT on its own the differences are more significant, here male PIs have an possible advantage over female PIs of $(40.7 \pm 117.4)\%$ in numbers, and 89.6% in time. Initially this was a surprise, and the reason was not obvious.

However, KPOT is "special" as it comprised of a small number of proposals, but with very substantial amounts of both requested and allocated observing time. We speculated that there could be "small number statistics" issues at play. To test this hypothesis we looked at what difference just one proposal being accepted or rejected could make. As shown in Table 6.7 the KPOT outcome was 19 accepted proposals with male PIs and 2 with female PIs, out of 54 and 8 submitted proposals, respectively. Now, imagine the situation where the outcome would have been 18 and 3 instead. This would have resulted in $(33.3 \pm 9.1)\%$ accepted proposals with male PIs, and $(37.5 \pm 25.4)\%$ with female PIs, a female advantage of 11.1% (rather than a male advantage of 40.7%). Regarding allocated time, the average amount of time allocated to male PIs was 263 hour per successful proposal and to female PIs less at 189 hour. Imagine that the hypothetical third successful female PI proposal would have been allocated 263 hour. This would have resulted in 4738 hour awarded to male PIs, or 30.1%, and 641 hour for female PIs, or 28.4%, a male advantage of 6.0% (rather than 89.6%), and in fact, smaller than the male advantage for OT1+OT2 p1 time (17.6%). This discussion shows that "small number statistics" are likely to be important in understanding the origin of the KPOT "outlier"; certainly the results are consistent with this hypothesis. At the same time, we are aware that this does not prove that this hypothesis necessarily is the correct explanation.

The above discussion still leaves the male advantages for the OT1 and OT2 AOs. Since the female proposer population is expected to be less senior than the male one and thus on average less experienced, we have investigated the experience of the populations of male and female proposer PIs. In order to determine the "academic age" of proposers we used the difference between the year that their PhD was awarded and the year that an AO to which they submitted a proposal was issued. Thus a proposer who submits multiple proposals to the same AO will be counted multiple times in the same "academic year", whilst one who submits multiple proposals to different AOs will be counted in different "academic years".

The year a proposer obtained their PhD (or equivalent) was determined for 96.4% of the proposals by searching the internet, particularly sites such as the Astrophysics Data Service (ADS), LinkedIn, the Astronomy Genealogy Project (astrogen.aas.org), ORCID.org, IEEE Xplore (https://ieeexplore.ieee.org/Xplore/home.jsp), and, for French theses https://www.theses.fr. Some proposers were contacted and provided their PhD dates by email. The proposers for which the PhD could not be found are often retired, deceased, or have left astronomy for some other reason. For PhD students who had not yet completed their degrees, the expected year of submission was used. For the small number of proposers who did not have a PhD and were not enrolled in a PhD programme, their dates were assumed to be arbitrarily far in the future. For some senior scientists in Italian institutes who do not have a PhD, a date three years after they obtained their Laurea was used. It should be noted that using the year of PhD to indicate the seniority does not take into account possible time spent outside of astronomy such as a career elsewhere.

Figure 6.2 shows the numbers of male and female Herschel PIs who obtained their PhDs or equivalent in one year bins between 10 years before the PhD date to 50 years after. Male PIs have a mean "academic age" of 13.5 years post-PhD compared to that of the female PIs of 8.9 years. The mean "academic age" of all PIs is 12.2 years. A two-sample Anderson-Darling Test (e.g., [104]) shows that the hypothesis that both samples come from the same underlying population can be rejected at >99% confidence.

We investigated the acceptance rates of proposals against PhD year. This is shown plotted for years −4 to 35 years after PhD in five year bins in Fig. 6.3. To the eye it may look like a slow increase in acceptance rates for above 10 years of academic age, however, the error bars are large and a least squares fit to the un-binned data gives an intercept of (52.5 ± 4.3) % and a gradient of (0.03 ± 0.29)% year^{-1} assuming that the uncertainties are the square root of the number of people in each bin. The value of R^2 is 0.42. The uncertainties are large and the gradient

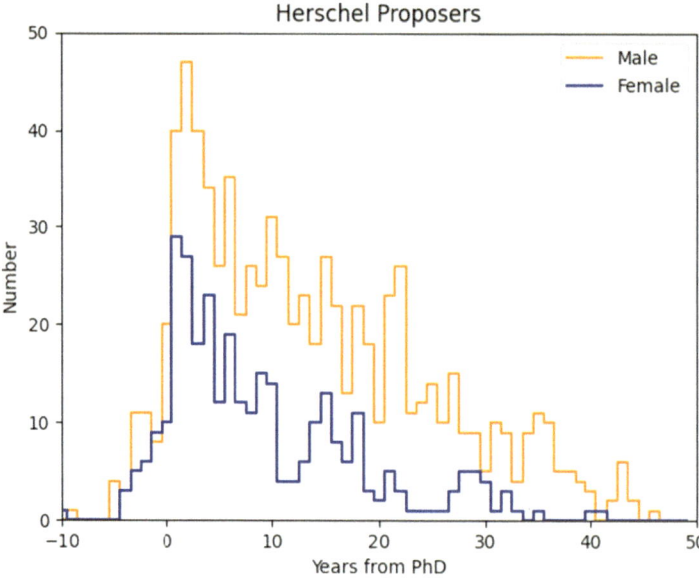

Fig. 6.2 The "academic age" (years since PhD) for male and female Herschel PIs. Male PIs have a mean "academic age" of 13.5 years compared to 8.9 years for female PIs

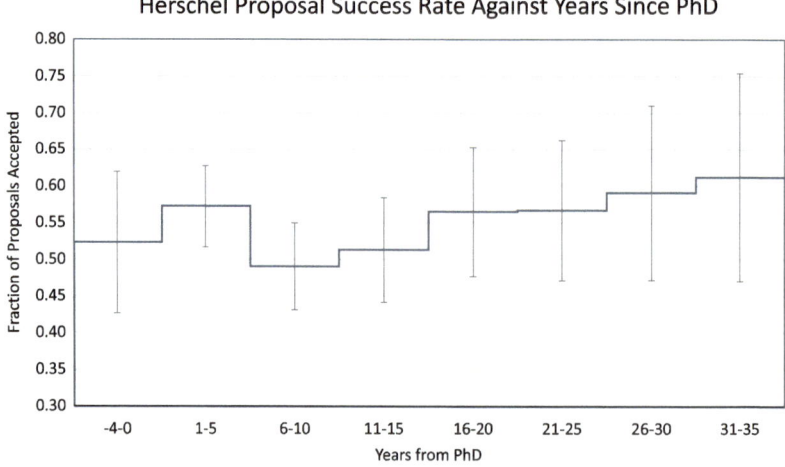

Fig. 6.3 The acceptance fraction of all Herschel PIs with year of their PhD. The indicative error bars indicate 1σ standard deviations assuming that the number of proposers in each age bin follows a Poisson distribution

is consistent with no change in proposal acceptance rate with "academic age". This is in contrast to the results from INTEGRAL and XMM-Newton which both show stronger evidence for increasing proposal success rates with "academic age".

6.10 Comparison with the Infrared Space Observatory (ISO)

The study of Herschel AOs and their outcomes is basically a snapshot from around the year 2010. This is in contrast to the ~20 year spans of AOs for XMM-Newton and INTEGRAL. However, as alluded to in Sect. 6.4, the Herschel community has a connection with earlier missions, notably the ESA ISO mission [105].

ISO predates the Horizon 2000 programme [2]. It was selected in 1983 based on a proposal from 1979, and launched on 17 November 1995. ISO was the world's first true orbiting infrared space observatory, providing astronomers with unprecedented sensitivity and observation capabilities for a detailed exploration of the universe at infrared wavelengths. Equipped with four state-of-the-art instruments, the ISO Infrared Camera (ISOCAM), the ISO Photo-polarimeter (ISOPHOT), the Short Wave Spectrometer (SWS) and the Long Wave Spectrometer (LWS) housed in a superfluid helium cryostat (the forerunner of the Herschel cryogenic system) it probed the sky in a range of photometric, polarimetric and spectroscopic modes in the spectral range 2.5–240 μm. The routine operational phase lasted until April 1998 during which time ISO made some 30,000 observations.

For ISO we do not have the same amount of data pertaining in particular to the statistics for proposals and observing time; at the time this was handled in a much more "manual" fashion compared to later missions. There were two AOs for observing proposals, featuring GT and OT as for other observatories, a pre-launch call in 1994, and a supplemental call in 1996. We could only find information on the total number of accepted proposals in the two calls which can be found in Table 6.8. Unfortunately, information on the number of submitted proposals could not be found.

The fraction of all accepted ISO proposals with a female PI is 15.0%. For the OT only it is 15.5% which can be compared with the same figure for Herschel (from Table 6.5) with a total of 644 OT proposals accepted out of which 183 had a female PI, or 28.4%. For the Herschel p1 proposals (only) the numbers are a total of 386 OT proposals accepted out of which 105 had a female PI, or 27.2%. This means that the fraction of successful female PIs increased by a factor of 1.83 (all) or 1.75 (p1

Table 6.8 The number of accepted proposals for the ISO GT and OT calls and the corresponding numbers with female PIs. Information on the number of submitted proposals is unfortunately missing. %(f)PI is the percentage of accepted proposals with a female PI

ISO GT			ISO OT			ISO total		
No.props	No.(f)PI	%(f)PI	No.props	No. (f)PI	%(f)PI	No. props	No. (f)PI	%(f)PI
168	21	12.5	911	141	15.5	1079	162	15.0

only) between around 1995 for ISO and about 15 years later for Herschel. This may reflect an increase in the fraction of female (infrared) astronomers in the community by about 80% in the time of Herschel compared to the time of ISO.

To test this hypothesis the fraction of female astronomers in the community over time are needed. We have had limited success finding this data and we extracted results between mid 2011 to late 2023 from the International Astronomical Union (IAU)'s membership statistics (Digital Public Data) (see Chap. 1). In mid-2011 the fraction of female IAU members was 17.5%. This is below the fraction of successful Herschel proposals with female PIs of ~28% in the preceding years which is a factor ~1.6 higher. However, without the fraction of female astronomers in the community in around 1995 it is not possible to compare the corresponding ISO number of ~15% with the fraction of female IAU members at the time. If the corresponding ISO number also would be a factor of ~1.6 higher it would have been tempting to ascribe the increase in the fraction of female PIs from ISO to Herschel simply resulting from the increase in the fraction of female astronomers overall. Aside from the fact that we lack the required numbers, this could of course be an overly simplistic explanation.

Independently, we have received information concerning the membership of the American Astronomical Society (AAS), also incomplete, but indicating that the fraction of female members may have increased by a factor approximately in the range 1.45–1.65 between 1995 and 2010. If the actual number is anywhere in this range it is thus somewhat smaller than the ~1.8 ratio of the fractions of successful proposals with female PIs in Herschel versus ISO. This would thus go part of the way but needing additional components to fully explain the evolution observed, such as e.g., an increase in the fraction of infrared/sub-millimetre astronomers in the total population of astronomers, which would not at all be unthinkable in this time-frame with all the activities in the field, in space and on the ground.

There continue to be ~30 refereed ISO publications per year and by the end of 2021 the total number is approaching 2100, another example of longevity. An interesting aspect of the ISO publication curve is what occurred in the "Herschel time", when the annual number of ISO publications approximately doubled in the early years of Herschel 2009–2011, and was elevated for another three years. This is hypothesised as a "Herschel effect", more people may have been using ISO data with respect to planning and proposing for Herschel, and in interpreting Herschel data. Furthermore, although not easily visible in the number of Herschel publications curve as an increase similar to that of ISO, several of the early Atacama Large Millimetre/sub-millimetre Array (ALMA) observations had Herschel connections, using Herschel data, and generating Herschel publications. It appears that "next generation" missions may enhance the value of the data from earlier missions, generating additional publications in the process.

6.11 Discussion

Herschel is a highly successful mission that scores well in various "success metrics" as already pointed out in Sect. 6.4. As described and shown in Fig. 6.1 Herschel's productivity in terms of publications began early in the mission. This is not something that "happened" but was due to careful preparations in order to be able to use the observatory efficiently, since its short lifetime was consumable (cryogen) limited, thus the onus was on efficient operations from the very beginning.

The introduction of the SDP was generally considered a big success; it validated the actual science performance in various respects, it provided real-life performance numbers in time already for the first in-flight AO and it provided material used for outreach.

As already pointed out Herschel provides only snapshots when it comes to studies of the proposal and time allocation processes and their evolution. There are two aspects that we find interesting. The first is that, if we neglect the KPOT selection, there is a probable male advantage of (-3–18%) in these processes (see Table 6.7, which is similar to that found for XMM-Newton and INTEGRAL (see Sect. 7.5). The second is, indeed, the very different situation for the KPOT selection which at first sight appears to have been strongly in favour of male PIs (Table 6.7). However, although it cannot be positively confirmed, it is possible (see the discussion in Sect. 6.9) that the outcome has been strongly influenced by "small number statistics".

Supplemented by what we have in terms of proposal statistics from ISO in around 1995 (about 15 years before Herschel), we find that the fraction of successful proposals with female PIs is about 80% higher for Herschel than for ISO. This appears to be a larger increase than the increase in the female fraction of the overall astronomical community in the same time, but more statistical data would be needed to affirm this tentative conclusion.

Acknowledgments We thank the Herschel HSC staff for their assistance. This research was supported by the International Space Science Institute (ISSI) in Bern, through the ISSI Working Group project 'The Scientific Performance of ESA's Science Missions'.

References

1. G.L. Pilbratt, J.R. Riedinger, T. Passvogel, et al., Herschel space observatory. An ESA facility for far-infrared and submillimetre astronomy. Astron. Astrophys. **518**, L1 (2010)
2. *European Space Science – Horizon 2000*. SP–1070 (European Space Agency, Paris, 1984)
3. B. Fleck, V. Domingo, A.I. Poland, The SOHO mission. Solar Phys. **162**(1) (1995)
4. C.P. Escoubet, M. Fehringer, M. Goldstein, Introduction: the cluster mission. Ann. Geophys. **19**, 1197–1200 (2001)
5. F. Jansen, D. Lumb, B. Altieri, et al., XMM-Newton observatory. I. The spacecraft and operations. Astron. Astrophys. **365**, L1–L6 (2001)
6. K.H. Glassmeier, H. Boehnhardt, D. Koschny, et al., The Rosetta mission: flying towards the origin of the solar system. Space Sci. Rev. **128**, 1 (2007)

7. V. Minier, R.M. Bonnet, V. Bontems, et al., *Inventing a Space Mission – The Story of the Herschel Space Observatory*. ISSI Scientific Report Series, vol. 14 (Springer, Berlin, 2017)

8. S. Clark, G. Pilbratt, *Herschel - Science and Legacy*, ed. by K. O'Flaherty, G. Pilbratt. ESA Brochure: ESA (2019)

9. J.A. Tauber, N. Mandolesi, J.L. Puget, et al., Planck pre-launch status: the Planck mission. Astron. Astrophys. **520**, A1 (2010)

10. D. Doyle, G. Pilbratt, J. Tauber, The Herschel and Planck space telescopes. Inst. Electr. Electron. Eng. **97**, 1043 (2009)

11. A. Poglitsch, C. Waelkens, N. Geis, et al., The photodetector array camera and spectrometer (PACS) on the Herschel space observatory. Astron. Astrophys. **518**, L2 (2010)

12. M.J. Griffin, A. Abergel, A. Abreu, et al., The Herschel-SPIRE instrument and its in-flight performance. Astron. Astrophys. **518**, L3 (2010)

13. T. de Graauw, F.P. Helmich, T.G. Phillips, et al., The Herschel-Heterodyne instrument for the far-infrared (HIFI). Astron. Astrophys. **518**, L6 (2010)

14. J. Riedinger, A First in astrophysics missions. The making of the Herschel science ground segment. ESA Bullet. **139**, 12 (2009)

15. S. Ott, The Herschel data processing system - HIPE and pipelines - up and running since the start of the mission, in *Astronomical Data Analysis Software and Systems XIX*, ed. by Y. Mizumoto, K.I. Morita, M. Ohishi. ASP Conference Series, vol. 434 (2010), p. 139

16. S. Ott, Herschel data processing development: 10 years after, in *Astronomical Data Analysis Software and Systems XXII*, ed. by D.N. Friedel. ASP Conference Series, vol. 475 (2013), p. 197

17. A. Wilson, (ed.). *The Far Infrared and Submillimetre Universe*. ESA SP-401. (ESA, Paris, 1997)

18. G.L. Pilbratt, J. Cernicharo, A.M. Heras, T. Prusti, R. Harris, (eds.) *The Promise of the Herschel Space Observatory*. ESA SP-460 (ESA, Paris, 2001)

19. E. Valiante, M.W.L. Smith, S. Eales, et al., The Herschel-ATLAS data release 1 - I. Maps, catalogues and number counts. Month. Not. R. Astron. Soc. **462**, 3146 (2016)

20. S.J. Maddox, E. Valiante, P. Cigan, et al., The Herschel-ATLAS data release 2. Paper II. Catalogs of far-infrared and submillimeter sources in the fields at the south and north galactic poles. Astrophys. J. Suppl. **236**, 30 (2018)

21. G. Rodighiero, E. Daddi, I. Baronchelli, et al., The Lesser role of starbursts in star formation at z=2. Astrophys. J. Lett. **739**, L40 (2011)

22. D. Elbaz, R. Leiton, N. Nagar, et al., Starbursts in and out of the star-formation main sequence. Astron. Astrophys. **616**, A110 (2018)

23. D. Elbaz, M. Dickinson, H.S. Hwang, et al., GOODS-Herschel: an infrared main sequence for star-forming galaxies. Astron. Astrophys **533**, A119 (2011)

24. S. Dye, L. Dunne, S. Eales, et al., Herschel-ATLAS: evolution of the 250 μm luminosity function out to z = 0.5. Astron. Astrophys. **518**, L10 (2010)

25. C. Gruppioni, F. Pozzi, G. Rodighiero, et al., The Herschel PEP/HerMES luminosity function - I. Probing the evolution of PACS selected Galaxies to z 4. Month. Not. R. Astron. Soc. **432**, 23 (2013)

26. L. Dunne, H.L. Gomez, E. da Cunha, et al., Herschel-ATLAS: rapid evolution of dust in galaxies over the last 5 billion years. Month. Not. R. Astron. Soc. **417**, 1510 (2011)

27. D. Lutz, Far-infrared surveys of galaxy evolution. Ann. Rev. Astron. Astrophys. **52**, 373 (2014)

28. D.A. Riechers, C.M. Bradford, D.L. Clements, et al., A dust-obscured massive maximum-starburst galaxy at a redshift of 6.34. Nature **496**, 329 (2013)

29. C.D. Dowell, A. Conley, J. Glenn, et al., HerMES: candidate high-redshift galaxies discovered with herschel/SPIRE. Astrophys. J. **780**, 75 (2014)

30. R.J. Ivison, A.J.R. Lewis, A. Weiss, et al., The space density of luminous dusty star-forming galaxies at z>4: SCUBA-2 and LABOCA imaging of ultrared galaxies from Herschel-ATLAS. Astrophys. J. **832**, 78 (2016)

31. D.A. Riechers, T.K.D. Leung, R.J. Ivison, et al., Rise of the titans: a dusty, hyper-luminous "870 μm riser" galaxy at z 6. Astrophys. J. **850**, 1 (2017)
32. J. Fischer, E. Sturm, E. González-Alfonso, et al., Herschel-PACS spectroscopic diagnostics of local ULIRGs: conditions and kinematics in Markarian 231. Astron. Astrophys. **518**, L41 (2010)
33. E. González-Alfonso, J. Fischer, J. Gracía-Carpio, et al., The Mrk 231 molecular outflow as seen in OH. Astron. Astrophys. **561**, A27 (2014)
34. E. Sturm, E. González-Alfonso, S. Veilleux, et al., Massive molecular outflows and negative feedback in ULIRGs observed by Herschel/PACS. Astrophys. J. Lett. **733**, L16 (2011)
35. M. Matsuura, E. Dwek, M. Meixner, et al., Herschel detects a massive dust reservoir in supernova 1987A. Science **333**, 1258 (2011)
36. M. Matsuura, E. Dwek, M.J. Barlow, et al., A stubbornly large mass of cold dust in the ejecta of supernova 1987A. Astrophys. J. **800**, 50 (2015)
37. M.J. Barlow, B.M. Swinyard, P.J. Owen, et al., Detection of a noble gas molecular Ion, 36ArH+, in the crab nebula. Science **342**, 1343 (2013)
38. P. André, A. Men'shchikov, S. Bontemps, et al., From filamentary clouds to prestellar cores to the stellar IMF: initial highlights from the Herschel Gould Belt survey. Astron. Astrophys. **518**, L102 (2010)
39. A. Men'shchikov, P. André, P. Didelon, et al., Filamentary structures and compact objects in the Aquila and Polaris clouds observed by Herschel. Astron. Astrophys. **518**, L103 (2010)
40. D. Arzoumanian, P. André, P. Didelon, et al., Characterizing interstellar filaments with Herschel in IC 5146. Astron. Astrophys. **529**, L6 (2011)
41. D. Arzoumanian, P. André, V. Könyves, et al., Characterizing the properties of nearby molecular filaments observed with Herschel. Astron. Astrophys. **621**, A42 (2019)
42. V. Könyves, P. André, A. Men'shchikov, et al., The Aquila prestellar core population revealed by Herschel. Astron. Astrophys. **518**, L106 (2010)
43. V. Könyves, P. André, A. Men'shchikov, et al., A census of dense cores in the Aquila cloud complex: SPIRE/PACS observations from the Herschel Gould Belt survey. Astron. Astrophys. **584**, A91 (2015)
44. V. Könyves, P. André, D. Arzoumanian, et al., Properties of the dense core population in Orion B as seen by the Herschel Gould Belt survey. Astron. Astrophys. **635**, A34 (2020)
45. D. Ward-Thompson, J.M. Kirk, P. André, et al., A Herschel study of the properties of starless cores in the Polaris Flare dark cloud region using PACS and SPIRE. Astron. Astrophys. **518**, L102 (2010)
46. N. Schneider, P. André, V. Könyves, et al., What determines the density structure of molecular clouds? A case study of Orion B with Herschel. Astrophys. J. Lett. **766**, L17 (2013)
47. V. Könyves, P. André, N. Schneider, et al., Growing evidence for a core formation threshold traced in Herschel Gould Belt survey clouds. Astron. Nachr. **334**, 908 (2013)
48. P. André, The Herschel view of star formation, in *XXVIIIth IAU General Assembly: Highlights of Astronomy*, ed. by T. Montmerle, vol 16 (University of Cambridge Press, Cambridge, 2015), p. 31
49. P. André, J. Di Francesco, D. Ward-Thompson, et al., From filamentary networks to dense cores in molecular clouds: toward a new paradigm for star formation, in *Protostars and Planets VI*, ed. by H. Beuther, R.S. Klessen, C.P. Dullemond, et al. The University of Arizona Press in collaboration with the Lunar and Planetary Institute (2014), p. 27
50. F. Motte, A. Zavagno, S. Bontemps, et al., Initial highlights of the HOBYS key program, the Herschel imaging survey of OB young stellar objects. Astron. Astrophys. **518**, L77 (2010)
51. A. Zavagno, D. Russeil, F. Motte, et al., Star formation triggered by the Galactic H II region RCW 120. First results from the Herschel Space Observatory. Astron. Astrophys. **518**, L81 (2010)
52. E.A. Bergin, T.G. Phillips, C. Comito, et al., Herschel observations of EXtraOrdinary sources (HEXOS): the present and future of spectral surveys with Herschel/HIFI. Astron. Astrophys. **521**, L20 (2010)

53. N.R. Crockett, E.A. Bergin, J.L. Neill, et al., Herschel observations of EXtraOrdinary sources: analysis of the HIFI 1.2 THz wide spectral survey toward Orion KL. I. Methods. Astrophys. J. **787**, 112 (2014)
54. N.R. Crockett, E.A. Bergin, J.L. Neill, et al., Herschel observations of EXtraOrdinary sources: analysis of the HIFI 1.2 THz wide spectral survey toward Orion KL. II. Chemical implications. Astrophys. J. **806**, 239 (2015)
55. S. Molinari, B. Swinyard, J. Bally, et al., Hi-GAL: the Herschel infrared galactic plane survey. Publ. Astron. Soc. Pac. **122**, 314 (2010)
56. S. Molinari, B. Swinyard, J. Bally, et al., Clouds, filaments, and protostars: the Herschel Hi-GAL milky way. Astron. Astrophys. **518**, L100 (2010)
57. S. Molinari, J. Bally, S. Glover, et al., The milky way as a star formation engine, in *Protostars and Planets VI*, ed. by H. Beuther, R.S. Klessen, C.P. Dullemond, et al., The University of Arizona Press in Collaboration with the Lunar and Planetary Institute (2014), p. 125
58. S. Molinari, E. Schisano, D. Elia, et al., Hi-GAL, the Herschel infrared galactic plane survey: photometric maps and compact source catalogues. Astron. Astrophys. **591**, A149 (2016)
59. D. Elia, S. Molinari, E. Schisano, et al., The Hi-GAL compact source catalogue - I. The physical properties of the clumps in the inner Galaxy (-71°.0 < l < 67°.0). Month. Not. R. Astron. Soc. **471**, 100 (2017)
60. D. Elia, M. Merello, S. Molinari, et al., The Hi-GAL compact source catalogue - II. The 360° catalogue of clump physical properties. Month. Not. R. Astron. Soc. **504**, 2742 (2021)
61. S. Molinari, A. Noriega-Crespo, J. Bally, et al., Large-scale latitude distortions of the inner Milky Way disk from the Herschel/Hi-GAL survey. Astron. Astrophys. **588**, A75 (2016)
62. D. Elia, F. Strafella, S. Dib, et al., Multifractal analysis of the interstellar medium: first application to Hi-GAL observations. Month. Not. R. Astron. Soc. **481**, 509 (2018)
63. W.D. Langer, T. Velusamy, J.L. Pineda, et al., C^+ detection of warm dark gas in diffuse clouds. Astron. Astrophys. **521**, L17 (2010)
64. J.L. Pineda, W.D. Langer, T. Velusamy, et al., A Herschel [C II] Galactic plane survey I. The global distribution of ISM gas components. Astron. Astrophys. **554**, A103 (2013)
65. W.D. Langer, T. Velusamy, J.L. Pineda, et al., A Herschel [C II] Galactic plane survey II. CO-dark H2 in clouds. Astron. Astrophys. **561**, A122 (2014)
66. J.L. Pineda, W.D. Langer, P.F. Goldsmith, et al., A Herschel [C II] Galactic plane survey III. [C II] as a tracer of star formation. Astron. Astrophys. **570**, A121 (2014)
67. C. Eiroa, D. Fedele, J. Maldonado, et al., Cold DUst around NEarby Stars (DUNES). First results. A resolved exo-Kuiper belt around the solar-like star ζ^2 Ret. Astron. Astrophys. **518**, L131 (2010)
68. B.C. Matthews, B. Sibthorpe, G. Kennedy, et al., Resolving debris discs in the farinfrared: early highlights from the DEBRIS survey. Astron. Astrophys. **518**, L135 (2010)
69. B. Sibthorpe, B. Vandenbussche, J.S. Greaves, et al., The vega debris disc: a view from Herschel. Astron. Astrophys. **518**, L130 (2010)
70. R. Liseau, C. Eiroa, D. Fedele, et al., Resolving the cold debris disc around a planet-hosting star. PACS photometric imaging observations of q1 Eridani (HD 10647, HR 506). Astron. Astrophys. **518**, L132 (2010)
71. B. Vandenbussche, B. Sibthorpe, B. Acke, et al., The β Pictoris disk imaged by Herschel PACS and SPIRE. Astron. Astrophys. **518**, L133 (2010)
72. C. Eiroa, J.P. Marshall, A. Mora, et al., DUst around NEarby stars. The survey observational results. Astron. Astrophys. **555**, A11 (2013)
73. B. Montesinos, C. Eiroa, A.V. Krivov, et al., Incidence of debris discs around FGK stars in the solar neighbourhood. Astron. Astrophys. **593**, A51 (2016)
74. B. Acke, M. Min, C. Dominik, et al., Herschel images of Fomalhaut. An extrasolar Kuiper belt at the height of its dynamical activity. Astron. Astrophys. **540**, A125 (2012)
75. B.L. de Vries, B. Acke, A.D.L. Blommaert, et al., Comet-like mineralogy of olivine crystals in an extrasolar proto-Kuiper belt. Nature **490**, 74 (2012)

76. L. Decin, N.L.J. Cox, P. Royer, et al., The enigmatic nature of the circumstellar envelope and bow shock surrounding Betelgeuse as revealed by Herschel. I. Evidence of clumps, multiple arcs, and a linear bar-like structure. Astron.Astrophys. **548**, A113 (2012)

77. L. Decin, M. Agúndez, M.J. Barlow, et al., Warm water vapour in the sooty outflow from a luminous carbon star. Nature **467**, 64 (2010)

78. L. Decin, Late stages of stellar evolution - Herschel's contributions. Adv. Space Res. **50**, 843 (2012)

79. E.F. van Dishoeck, E.A. Bergin, D.C. Lis, et al., Water: from clouds to planets, in *Protostars and Planets VI*, ed. by H. Beuther, R.S. Klessen, C.P. Dullemond, et al. The University of Arizona Press in Collaboration with the Lunar and Planetary Institute (2014), p. 835

80. P. Caselli, E. Keto, E.A. Bergin, et al., First detection of water vapour in a Prestellar core. Astrophys. J. Lett. **759**, L37 (2012)

81. L.E. Kristensen, E.F. van Dishoeck, E.A. Bergin, et al., Water in star-forming regions with Herschel (WISH). II. Evolution of 557 GHz 110-101 emission in low-mass protostars. Astron. Astrophys. **542**, A8 (2012)

82. M.R. Hogerheijde, E.A. Bergin, C. Brinch, et al., Detection of the water reservoir in a forming planetary system. Science **334**, 338 (2011)

83. E.A. Bergin, L.I. Cleeves, U. Gorti, et al., An old disk still capable of forming a planetary system. Nature **493**, 644 (2013)

84. T. Cavalié, H. Feuchtgruber, E. Lellouch, et al., Spatial distribution of water in the stratosphere of Jupiter from Herschel HIFI and PACS observations. Astron. Astrophys. **553**, A21 (2013)

85. P. Hartogh, E. Lellouch, R. Moreno, et al., Direct detection of the Enceladus water torus with Herschel. Astron. Astrophys. **532**, L2 (2011)

86. T. Cavalié, V. Hue, P. Hartogh, et al., Herschel map of Saturn's stratospheric water, delivered by the plumes of Enceladus. Astron. Astrophys. **630**, A87 (2019)

87. M. Küppers, L. O'Rourke, D. Bockelée-Morvan, et al., Localized sources of water vapour on the dwarf planet (1) Ceres. Nature **505**, 525 (2014)

88. P. Hartogh, J. Crovisier, M. de Val-Borro, et al., HIFI observations of water in the atmosphere of comet C/2008 Q3 (Garradd). Astron. Astrophys. **518**, L150 (2010)

89. P. Hartogh, D. Lis, D. Bockelée-Morvan, et al., Ocean-like water in the Jupiterfamily comet 103P/Hartley 2. Nature **478**, 218 (2011)

90. D. Bockelée-Morvan, N. Biver, B. Swinyard, et al., Herschel measurements of the D/H and 16O/18O ratios in water in the Oort-cloud comet C/2009 P1 (Garradd). Astron. Astrophys. **544**, L15 (2012)

91. D.C. Lis, N. Biver, D. Bockelée-Morvan, et al., A Herschel study of D/H in water in the Jupiter-family Comet 45P/Honda-Mrkos-Pajdušáková and prospects for D/H measurements with CCAT. Astrophys. J. Lett. **774**, L3 (2013)

92. K. Altwegg, H. Balsiger, A. Bar-Nun, et al., 67P/Churyumov-Gerasimenko, a Jupiter family comet with a high D/H ratio. Science **347**, 1261952 (2015)

93. E. Vilenius, C. Kiss, M. Mommert, et al., "TNOs are cool": a survey of the trans-Neptunian region. VI. Herschel/PACS observations and thermal modeling of 19 classical Kuiper belt objects. Astron. Astrophys. **541**, A94 (2012)

94. E. Vilenius, C. Kiss, T. Müller, et al., "TNOs are cool": a survey of the trans-Neptunian region. X. Analysis of classical Kuiper belt objects from Herschel and Spitzer observations. Astron. Astrophys. **564**, A35 (2014)

95. P. Lacerda, S. Fornasier, E. Lellouch, et al., The Albedo-Color diversity of transneptunian objects. Astrophys. J. Lett. **793**, L2 (2014)

96. E. Lellouch, P. Santos-Sanz, S. Fornasier, et al., The long-wavelength thermal emission of the Pluto-Charon system from Herschel observations. Evidence for emissivity effects. Astron. Astrophys. **588**, A2 (2016)

97. Astron. Astrophys. **518**, L1–L152 (2010)

98. Astron. Astrophys. **521**, L1–L50 (2010)

99. R. Shirley, K. Duncan, M.C. Campos Varillas, et al., HELP: the Herschel extragalactic legacy project. Month. Not. R. Astron. Soc. **507**, 129 (2021)

100. E. Sciacca, D. Vitello, U. Becciani, et al., VIALACTEA science gateway for Milky Way analysis. Future Gen. Comput. Syst. **94**, 947 (2021)
101. J.I. Davies, M. Baes, S. Bianchi, et al., DustPedia: a definitive study of cosmic dust in the local universe. Publ. Astr. Soc. Pac. **129**, 044102 (2017)
102. T.G. Müller, A. Marciniak, C. Kiss, et al., Small bodies near and far (SBNAF): a benchmark study on physical and thermal properties of small bodies in the Solar System. Adv. Space Res. **62**, 2326 (2018)
103. I.N. Reid, Gender-correlated systematics in HST proposal selection. Publ. Astron. Soc. Pac. **126**(944), 923 (2014)
104. G.J. Babu, E.D. Feigelson, Goodness-of-fit and all that! ASP Conf. **351**, 127 (2006)
105. M.F. Kessler, J.A. Steinz, M.E. Anderegg, et al., The infrared space observatory (ISO) mission. Astron. Astrophys. **315**(2), L27 (1996)

Chapter 7
Conclusions

Arvind Parmar, Roger-Maurice Bonnet, Guido De Marchi, Erik Kuulkers, Göran Pilbratt, Norbert Schartel, and John Zarnecki

Abstract This chapter summarises and draws together the results of the previous chapters of this work. Using the assumption that the exploitation and contributions to the Science Programme should scale with a Member State's financial contribution to the programme, the "exploitation" (number of publications) is contrasted with the "contributions" (number of payload elements provided) to the ESA Science Programme. Some of the ESA Member States who excel in these areas are highlighted. The acceptance rates of proposals submitted requesting observing time on ESA's observatories (XMM-Newton, INTEGRAL and Herschel) are used to investigate success rate dependence on academic age, gender and institute country.

Keywords ESA Science Programme · Scientific literature · Science productivity · Peer review · ESA Member State exploitation and contributions · Observing time outcomes · Academic age · Institute countries · Gender · Ideas for future studies · XMM-Newton · INTEGRAL · Herschel

A. Parmar (✉)
Department of Space and Climate Physics, MSSL/UCL, Dorking, UK
e-mail: arvind.parmar@ucl.ac.uk

R.-M. Bonnet
Institut d'Astrophysique de Paris, Paris, France
e-mail: rmbonnet@issibern.ch

G. De Marchi · E. Kuulkers
Directorate of Science, ESA/ESTEC, Noordwijk, The Netherlands
e-mail: gdemarchi@esa.int; Erik.kuulkers@esa.int

G. Pilbratt
Göteborg, Sweden
e-mail: glpil@pm.me

N. Schartel
Directorate of Science, ESA/ESAC, Villanueva de la Cañada, Madrid, Spain
e-mail: norbert.schartel@esa.int

J. Zarnecki
The Open University, Milton Keynes, UK
e-mail: J.C.Zarnecki@open.ac.uk

© The Author(s) 2025   183
A. Parmar et al., *ESA Science Programme Missions*, ISSI Scientific Report Series 18,
https://doi.org/10.1007/978-3-031-69004-4_7

7.1   Introduction

We have examined the performance of, and contributions to, the European Space Agency (ESA) Science Programme's missions. A brief description of the Programme is to be found in Chap. 1. Here we summarise the results from individual chapters and combine them to allow comparisons. We used:

- The number of refereed publications that have resulted from the data obtained by the ESA Science Programme's missions (Chap. 2).
- The number of payload Principal Investigator (PI)s and Co-Principal Investigator (co-PI) which we use as a proxy for the financial contributions by the ESA Member States to the payloads and scientific ground segment elements (Chap. 3).
- The number of observing proposals submitted and accepted to the ESA Science Programme's observatory missions, XMM-Newton, INTEGRAL and Herschel (Chaps. 4, 5 and 6).

There are a number of other areas where Member States may directly benefit from the ESA Science Programme such as industrial return and the development of innovative technologies which are outside the scope of this assessment.

7.2   Refereed Publications

We counted the numbers of refereed publications from scientists located in the different ESA Member States. We assumed that the number of publications from a Member State should scale with their financial contributions to the Programme. We then used the number of publications from scientists located in a Member State divided by the 2000–2021 financial contribution from that Member State as a metric to measure of how well a Member State is exploiting the scientific opportunities resulting from the ESA Science Programme's missions.

Using this metric we investigated publication numbers from the four Member States who make the largest financial contributions to the ESA Science Programme as these have sufficient numbers of PIs and co-PIs to make the most reliable comparisons. Recalling that for publications, the four ESA Member States that contribute the most to the ESA Programme (France, Germany, Italy and the United Kingdom) have first author publications of -4, -14, +23, and +23%, respectively, when compared to their 2000–2021 financial contributions (Chap. 2). For the smaller ESA Member States it is notable that the Netherlands has 42% more publications than expected given its contribution and Finland has similarly 40% more than predicted. Not all of the smaller ESA Member States have more first author publications than expected with Norway, Romania and Portugal all having $\leq 30\%$ of the expected numbers of publications.

We also examined the publication statistics using all the authors on a publication and all the authors normalised so that each publication contributes exactly one author. The latter is our preferred method of measuring the overall contribution of countries, whereas the number of first authors provides a better measure of "leadership" in scientific exploitation. The above conclusions are not changed significantly if the all-author metrics are used instead of the first authors.

7.3 Payload Contributions

The contributions to the payloads of 28 ESA-led missions were examined using the numbers of payload PIs and co-PIs as a proxy. A total of 337 originally selected PIs and co-PIs were found of which 279 were located in the ESA Member States. The four ESA Member States that contribute the most to the Science Programme (Germany, the United Kingdom, France and Italy) dominate the provision of payloads having 192 original PIs and co-PIs which is 68% of the total. The unweighted payload contributions, compared to the financial contributions to the 2000–2011 contributions to the ESA Science Programme, are +27%, +5%, +20%, and −12% for France, Germany, Italy and the United Kingdom, respectively. For these larger ESA Member States, the outcomes are not significantly changed when the payload contributions are weighted by factors to take into account (1) mission class, (2) the number of PIs and co-PIs on a mission and (3) the relative contributions of the average PI compared to the average co-PI. This lack of change suggests a relatively robust process, at least for the results for the larger Member States. We note that all the ESA Member States except for Greece, Luxembourg and Romania have PIs or co-PIs with Estonia having its first co-PI in 2019 and Portugal in 2020.

We also examined the genders and "academic ages" of the PIs and co-PIs who provided payload elements. 9.5% of the originally appointed PIs and co-PIs were female. We determined the average difference between the year that the payload of a mission was approved and the year in which each originally selected PI and co-PI received their PhD to be 15.9 years (Table 7.1). Since being a PI or co-PI on an instrument can be considered to be a leadership position we would expect that

Table 7.1 The mean "academic age" (the difference in years between when a scientist was awarded their PhD and the year that an AO to which they submitted a proposal was issued) for observing time PIs for XMM-Newton, INTEGRAL and Herschel. For INTEGRAL only proposals that requested observing time were included. The "academic ages" of the original PIs for payloads for the ESA Science Programme missions are also shown for comparison. C is the completeness of the PhD information as a percentage

Mission	C	Mean "Academic Age" (years)		
		Total	Male	Female
XMM-Newton	94.8	10.9	11.2	9.8
INTEGRAL	99.7	13.7	14.7	10.8
Herschel	96.4	12.2	13.5	8.9
Payloads	84.2	15.9	15.7	17.9

this community consists of scientists with considerable experience. To check this, assuming an average age for obtaining a PhD of 25–30 years, implies an average age of 40–45 years. The current (1 January 2023) International Astronomical Union (IAU) membership for this age range comprises around 26% females. This is similar to the percentage of female PIs and co-PIs in recent payload selections (∼25%) suggesting comparable selection chances for male and female PIs.

7.4 Publications and Payloads

We investigated whether there are any correlations between Member States' payload contributions and their exploitation of the scientific data from the Programme's missions. Figure 7.1 shows the fraction of original PIs and co-PIs from each ESA Member State divided by the 2000–2021 financial contributions plotted against the similarly normalised number of first authored publications. The plotted uncertainties on the relative payload contributions are indicative only and obtained using the square roots of the number of PIs and co-PIs for each Member State.

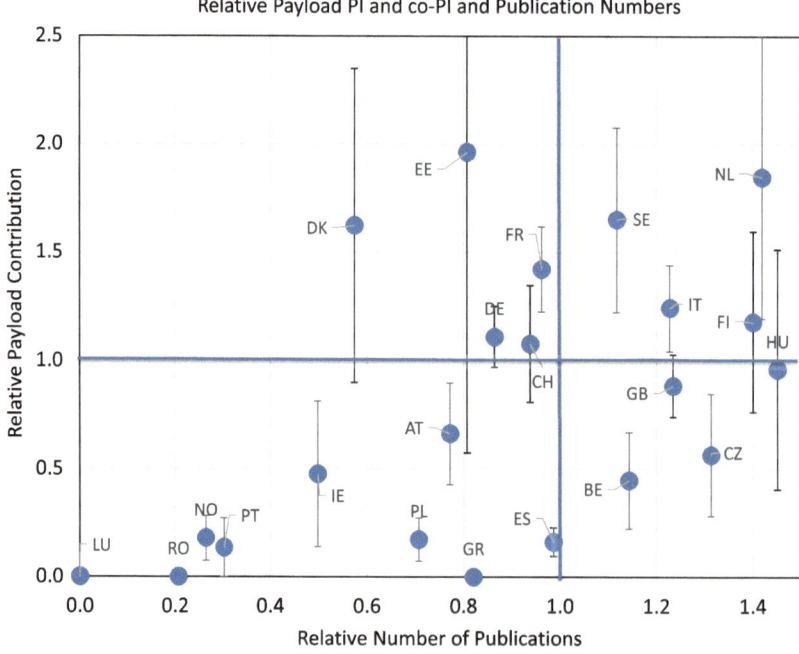

Fig. 7.1 The fraction of original PIs and co-PIs from each ESA Member State divided by the 2000–2021 financial contributions plotted against the similarly normalised number of first authored publications. The error bars are indicative and assume Poisson statistics apply to the numbers of PIs and co-PIs

The uncertainties on the publication numbers are much smaller due to their far higher number (22,930) compared to PIs and co-PIs (279). The meeting point of the horizontal and vertical lines at (1,1) is the position where a Member State whose exploitation as measured by first author publications and contribution to payloads as measured by the number of PIs and co-PIs is in balance. It is notable that the big four Member States cluster around this position. Sweden, Italy, the Netherlands and Finland are located in the top-right quadrant and so excel in both exploitation and in their payload contributions. Combining the payload and first author publication values shows that:

- Italy plays an important role in the ESA Science Programme being a strong contributor to payloads (+20% compared to expectation) whilst also being strong in scientific exploitation, obtaining +23% more first author publications than expected.
- France is also an important contributor to payloads, being the largest relative contributor (+27% compared to expectation), but is not so strong in science exploitation, obtaining 4% less first author publications than expected.
- Germany is similar to France in that it provides 5% more payload elements than predicted on the basis of the relative financial contributions, but obtains 14% fewer first author publications than predicted.
- The United Kingdom has different strengths to France, providing 12% fewer payload elements than expected, but being good at scientific exploitation, having 23% more first author publications than expected on the basis of the relative financial contributions.

7.5   Observing Time Proposals

We have examined the number of submitted and accepted proposals and the amount of awarded observing time for three of the ESA Science Programme's observatories. It is likely that a selection process that provides equitable outcomes for male and female PIs will likely provide for the best overall scientific outcomes. We are also aware that a mission which does not provide opportunities for early career scientists may not be sustainable in the long term. Of the three missions evaluated, XMM-Newton has by far the best statistics with 10,579 Open Time proposals compared to 1508 requesting observing time for INTEGRAL and 1170 for Herschel. Both INTEGRAL and XMM-Newton proposals cover intervals of over 20 years allowing the temporal evolution of their user communities to be investigated. Both missions show increasing fractions of proposals led by female PIs with time, which presumably reflects the increasing fraction of female astronomers in the community over the last 20 years.

Table 7.1 summarises the mean "academic ages" (the difference in years between when a scientist was awarded their PhD and the year that an AO to which they submitted a proposal was issued) for the PIs requesting observing time on XMM-

Table 7.2 The acceptance rate of observing time proposals against "academic age" showing evidence that the acceptance rate increases with seniority. Acceptance assumes that any amount of observing time was awarded to a proposal. "academic age" is the difference in years between when a scientist was awarded their PhD and the year that an AO to which they submitted a proposal was issued

Mission	Intercept (Percent)	Gradient (% Year^{-1})
XMM-Newton	37.3 ± 0.4	0.35 ± 0.15
INTEGRAL	45.7 ± 0.4	0.73 ± 0.26
Herschel	52.5 ± 4.3	0.03 ± 0.29

Newton, INTEGRAL and Herschel. The "academic ages" are given for all PIs and for female and male PIs separately. For all three missions, the population of female PIs is younger than for their male colleagues. It can be seen that XMM-Newton has a significantly younger population of PIs than the other two observatories. We also show in Table 7.1 the same information for the PIs of the payload provision for the ESA Science Programme's missions which, as discussed above, have significantly higher mean ages.

We have also used the number of years since PhD or "academic age" to show that for XMM-Newton and INTEGRAL proposal success rates probably increase with "academic age" whereas this may not be the case for Herschel (Table 7.2).

The proposal success rates for male and female astronomers were evaluated in a number of different ways and the range of outcomes were used to estimate the overall success rates:

- For XMM-Newton AO-2 to AO-21 proposals, male PIs appear to be 5–15% more successful than their female counterparts.
- For INTEGRAL AO-1 to AO-19 proposals, male PIs appear to be 2–11% more successful than their female counterparts.
- For Herschel OT1+OT2 proposals, male PIs appear to be −3–18% more successful than their female counterparts. The Key Programme Open Time (KPOT) appears to be an "outlier"; see Table 6.7 and discussion in Sect. 6.9.

From the three observatories studied, male PIs are likely to be more successful than female PIs in response to ESA AOs. The male to female success ratios measured in this study are closer to parity than for some other comparable missions and facilities and are similar to those from Hubble Space Telescope (HST) once this mission implemented dual-anonymous reviewing. This supports the view that the ESA proposal selection processes are robust and produce outcomes consistent with, or at least very close to, the best available science.

There does not appear to be an obvious correlation between the gender success rates with XMM-Newton and INTEGRAL and the fraction of female members of the relevant time allocation committees. For XMM-Newton, a higher fraction of female Observation Time Allocation Committee (OTAC) panel chairs may lead to more female-led proposals being accepted. This does not appear to be the case for

INTEGRAL, although the number of proposals is far fewer here. For Herschel, the fractions of female Time Allocation Committee (TAC) members remained the same during the three AOs so it is not possible to investigate any dependence with success rate.

We note that early career astronomers submitting proposals to ESA's observatories have success rates not too dissimilar to their more senior colleagues (see Table 7.2). The average proposal success rates for any grade for new PhDs are 37.3, 45.7, and 52.5% for XMM-Newton, INTEGRAL and Herschel, respectively. Colleagues who have had their PhDs for 20 years have success rates (7.2 $\pm$ 3.0), (14.6 $\pm$ 5.2), and (0.8 $\pm$ 5.8)% higher for XMM-Newton, INTEGRAL and Herschel, respectively (the uncertainties assume Poisson statistics). The Herschel results are consistent with there being no change in success rate with age. Given the 20 years that the more senior colleagues have had to develop networks and build up expertise these differences seem relatively small. Of course, some of this experience may be used in mentoring more junior colleagues and so contributing to their successes.

7.6 Final Considerations

We have demonstrated that ESA Science Programme missions remain highly scientifically productive for many years ($\gtrsim$20 years) after the end of operations. This emphasises the importance of maintaining the long-term archives of the ESA Science Programme's missions to ensure that the data can be scientifically exploited by a wide community. Their importance is enhanced as by this time the original instrument and Science Operations Centre (SOC) teams, and their expertise, are not always available. Well documented archives containing the relevant scientific products allow new generations of scientists to access "old" data. Thus we believe that it is important that ESA continues to support the exploitation of data from the archives, particularly for scientists located in the smaller ESA Member States who may not have the national infrastructure to support such exploitation.

It would be very useful for future studies similar to this if prospective instrument teams include the genders of their members, including the PI, type of position held (student, post-doctoral, faculty, or similar) as well as year of PhD in their proposals. ESA and the ESA Member States funding agencies can then easily monitor the gender distributions to ensure equitable outcomes. We expect that the number of female instrument team PIs will grow further with time given the increasing fraction of early career female astronomers who are now beginning to reach senior positions.

From discussions with many instrument team scientists we are aware that being a payload PI or co-PI is a very challenging task which typically lasts for a significant fraction of a career. It would be helpful if ESA and the national funding agencies established a mentoring system for new and prospective instrument PIs which would help guide less senior scientists in this often daunting task. ESA could also help train the next generation of payload investigators providing guidance on how to fulfil the responsibilities of the role, perhaps through a mentoring role of senior ESA payload engineers. Dedicated support for instrument PIs from smaller Member States who may not have the same amount of national infrastructure support as those from larger Member States would be especially helpful.

Future ESA AOs for observing time should request the gender and type of position held (student, post-doctoral, faculty, or similar) as well as year of PhD from PIs. ESA could then monitor the gender and age distributions of the outcomes of future observatory AOs to help ensure equitable outcomes. We suggest that this information remains confidential and not made available to reviewers so as not to bias selections. From our analysis, it is unclear how the composition and leadership of TACs affects the proposal selection process outcomes. An investigation comparing the results of different representative TACs evaluating the same set of proposals would be useful. However, given the amount of work involved in a typical TAC evaluation, this is a major challenge.

It would be interesting to investigate the reason, or reasons, why male PIs have higher proposal success rates than their female counterparts. There could be many factors including geographic. We examined regional dependencies for Europe, North America and the rest of the World. We found that Europe and North America are broadly similar in terms of success rates and gender balance, whereas the rest of the World is not with a significantly lower acceptance rate, but better gender balance. Dividing Europe into regions may provide further insights into the causes of the different success rates and strategies to overcome these.

Acknowledgments We thank Gaitee Hussain for her comments on an earlier version of this work. We thank Dominique Detain, Xiaolong Dong of ISSI Beijing and Tianran Sun and Yingjie Zhang of the National Space Science Center, Chinese Academy of Sciences, for their contributions. The International Space Science Institute (ISSI) Bern staff are thanked for their hospitality and assistance. In addition, Maurizio Falanga and Mark Sargent are acknowledged for their invaluable support and advice throughout this project. This research was supported by the International Space Science Institute (ISSI) in Bern, through the ISSI Working Group project 'The Scientific Performance of ESA's Science Missions'.

Index

A

Academic age, 84, 85, 102–104, 106, 111, 115, 140, 141, 145–148, 150, 171, 173–175, 185, 187, 188

Announcement of Observing Opportunities, 56, 92, 128

Archival papers, 53, 63–66

C

Co-Principal Investigator (co-PI), 73, 74, 76–85, 184–187, 189

Cosmic Vision, 2, 5

D

Dual-anonymous reviews, 116–118, 150, 151, 188

E

ESA Member State exploitation and contributions, 2, 17–19, 21, 25–29, 66, 74, 78–84, 184–186, 189

ESA Science Advisory Structure, 72, 77

ESA Science Programme, 1–7, 10, 13–15, 21, 26–28, 64, 66, 67, 71–85, 184, 185, 187–189

European Space Agency (ESA), 1, 10, 71, 96, 124, 156, 184

F

Further activities, 118–119

G

Gender, 5, 6, 73, 76, 82–85, 95–97, 100, 101, 105–107, 112, 114–118, 132, 134, 136, 137, 140, 141, 143, 144, 147, 150, 151, 167, 169–171, 185, 188–190

Gender balance, 5, 7, 149, 190

H

Herschel, 2, 5, 12, 17–19, 22–24, 26, 30, 31, 34–36, 41, 45, 50, 57–59, 72, 75, 82, 84, 155–177, 184, 185, 187–189

Horizon 2000, 2, 156, 175

I

Improvements for future studies, 189

Institute countries, 77, 97, 134

INTEGRAL, 5, 12, 17–19, 22, 23, 30, 31, 34, 41, 44, 72, 74, 82, 84, 93, 123–151, 163, 164, 170, 175, 177, 184, 185, 187–189

ISO comparison, 175–176

J

Joint programmes, 93, 130

L

Literature citations, 30, 32

Literature impact factors, 9–67

Literature statistics, 10, 16, 27, 31, 32, 34, 53

M

Member State, 1–3, 16–32, 66, 72–74, 76–85, 96, 97, 99, 135, 168, 184–187, 189, 190
Member State payload provision, 71–85
Mission selection, 4
Mission study phases, 4, 5

O

Observing priorities, 93, 94, 105, 110, 112, 115, 131, 138, 143–149, 165, 168
Observing time outcomes, 5, 148
Open Time, 92, 128, 130, 131, 165, 187
Over-subscription, 93, 94, 113, 128, 132, 140

P

Panel chairs, 94–96, 107, 110, 114, 129, 132–134, 142, 149, 167, 188
Panel members, 92, 95, 114, 132, 167
Payload complexity, 80–82
Peer review, 92, 108, 128, 143
Principal investigator (PI), 56, 57, 73, 74, 76, 77, 80–85, 88, 93, 96, 100–102, 105–107, 111–116, 118, 134, 135, 137, 139–141, 143–145, 147–151, 168, 170–173, 175, 184, 185, 189
Proposal acceptance rates, 106, 107, 114, 140–144, 148–150, 175

Proposer academic age, 102–104
Proposers' institutes countries, 97–100, 135–136

R

Regional dependence, 115–116

S

Science productivity, 163–164
Science results, 16, 88–91, 125–127
Scientific literature, 67, 73

T

Time Allocation Committee (TAC), 129–134, 138, 141–143, 149, 150, 188, 189, 190

U

User group, 5, 96–97

X

XMM-Newton, 2, 11, 12, 14, 17–19, 22–24, 30, 31, 33, 35, 36, 41, 42, 45, 57–59, 63, 65, 72, 74, 83, 84, 87–119, 127, 130, 134, 140, 147, 150, 156, 164, 175, 177, 184, 185, 187–189